Fisheries Management in Crisis

Also of interest

Aquaculture Development – Progress and Prospects
T.V.R. Pillay
0 85238 218 9

Aquaculture and the Environment
T.V.R. Pillay
0 85238 183 2

Commercial Fishing Methods
Third Edition
John C. Sainsbury
0 85238 217 0

The Common Fisheries Policy
Michael J. Holden
0 85238 205 7

The Economics of Salmon Aquaculture
Trond Bjørndal
0 632 02704 5

Exploitable Marine Ecosystems: Their Behaviour and Management
T. Laevastu, D.L. Alverson and R. Marasco
0 85238 225 1

Fisheries Biology, Assessment and Management
Michael King
0 85238 231 6

The Icelandic Fisheries: Evolution and Management of a Fishing Industry
Ragnar Arnason
0 85238 210 3

Large Marine Ecosystems: Stress, Mitigation and Sustainability
Edited by Kenneth Sherman, Lewis M. Alexander and Barry D. Gold
0 87168 506 X

Large Marine Ecosystem Sustainability: The Northeast Shelf
D.K. Sherman, N.A. Jaworski and T. Smayda
0 86542 468 3

FISHERIES MANAGEMENT IN CRISIS

EDITED BY

KEVIN CREAN AND DAVID SYMES

Fishing News Books

Copyright © 1996 by
Fishing News Books
A division of Blackwell Science Ltd
Editorial Offices:
Osney Mead, Oxford OX2 0EL
25 John Street, London WC1N 2BL
23 Ainslie Place, Edinburgh EH3 6AJ
238 Main Street, Cambridge,
   Massachusetts 02142, USA
54 University Street, Carlton,
   Victoria 3053, Australia

Other Editorial Offices:
Arnette Blackwell SA
   224, Boulevard Saint Germain
   75007 Paris, France

Blackwell Wissenschafts-Verlag GmbH
   Kurfürstendamm 57
   10707 Berlin, Germany

   Zehetnergasse 6
   A-1140 Wien
   Austria

First published 1996

Set in 9.5pt Souvenir Light
by DP Photosetting, Aylesbury, Bucks
Printed and bound in Great Britain
by Hartnolls Ltd, Bodmin, Cornwall

DISTRIBUTORS

Marston Book Services Ltd
PO Box 269
Abingdon
Oxon OX14 4YN
(Orders: Tel: 01865 206206
         Fax: 01865 721205
         Telex: 83355 MEDBOK G)

USA
   Blackwell Science, Inc.
   238 Main Street
   Cambridge, MA 02142
   (Orders: Tel: 800 215-1000
            617 876-7000
            Fax: 617 492-5263)

Canada
   Copp Clark, Ltd
   2775 Matheson Blvd East
   Mississauga, Ontario
   Canada, L4W 4P7
   (Orders: Tel: 800 263-4374
            905 238-6074)

Australia
   Blackwell Science Pty Ltd
   54 University Street
   Carlton, Victoria 3053
   (Orders: Tel: 03 9347-0300
            Fax: 03 9349-3016)

A catalogue record for this book is available
from the British Library

ISBN 0–85238–231–6

Library of Congress
Cataloging-in-Publication Data

Fisheries management in crisis/edited by Kevin
   Crean and David Symes.
       p. cm.
    Some of these papers were originally
   presented at a workshop held in May 1994 in
   Brussels.
    Includes bibliographical references (p.   )
   and index.
    ISBN 0-85238-231-6
    1. Fishery policy—Congresses. 2. Fishery
   management—Social aspects—Congresses.
   3. Fishery management—Economic aspects—
   Congresses. I. Crean, Kevin.   II. Symes,
   David.
   SH328.F563 1996
   333.95'615—dc20   95-51758
                                          CIP

Contents

v

Preface

The social science of fisheries management is a relatively new and, so far, uncoordinated field of study. However, its role as an applied social science is growing as policy makers discover the relevance of 'socio-economic factors' in determining the success or failure of their regulatory systems. To date, the analysis of the socio-economic factors has been left very largely to fisheries economists with a consequent emphasis upon measurable indices and little attempt to explore the underlying social issues and cultural values.

The present volume has its origins in a two-day workshop held in Brussels in May 1994, under the title of 'An Agenda for Social Science Research in Fisheries Management' and financed through a CEC sponsorship grant from the AIR Programme as part of the European Union's Third Programme Framework. The workshop was attended by 27 social scientists from 11 countries drawn from EU member states, associated countries and beyond, with observers from the Commission of the European Communities (Directorates General VI, VII and XIV), the European Parliament, the Organisation for Economic Cooperation and Development (OECD) and the United Nations' Food and Agriculture Organisation (FAO). From the 25 papers presented at the workshop, 16 have been selected for inclusion in this book. They reflect the five themes that structured the proceedings of the workshop: defining the social objectives of a fisheries policy; institutional restructuring for policy formulation and implementation; social issues in the fishing community; socio-economic development of fishery dependent regions; and the geopolitics of fish supplies, consumption and user/producer relations. Only the first and last chapters have been added.

Editing a set of papers drawn from several different countries and a wide range of disciplines can be a daunting task. This proved not to be the case and our first expression of appreciation must go to all the contributors who unfailingly met our deadlines and scarcely flinched at the way in which we rephrased their English. We must also express our thanks to Dominique Levieil of DG XIV who initiated the idea, to Emma Burnett who retyped many of the papers, and to Keith Scurr who redrew the maps and diagrams for publication. Our final acknowledgement is to Jeremy Phillipson who worked tirelessly behind the scenes at the workshop and in the preparation of the book.

Although the original workshop was undertaken with financial assistance from the Commission of the European Communities, the content of this book does not necessarily reflect the views of the Commission and in no way anticipates the Commission's future policy in this area.

Kevin Crean and *David Symes*
Hull

Contributors

Peter Friis
North Atlantic Regional Studies
Roskilde University
PO Box 260
DK-4000 Roskilde
Denmark

Torben A. Vestergaard
Department of Anthropology
Aarhus University
Moesgaard
DK-8270
Højbjerg
Denmark

Jógvan Mørkøre
Institute of History and Social Science
University of the Faroe Islands
J.C. Svaboesgøtu 7
Fr-100 Tórshavn
Faroe Islands

Geneviève Delbos
Corn er Porh
Locoal-Mendon
56 550 Belz
France

Gérard Prémel
23 bd de Lattre de Tassigny
35000 Rennes
France

Annie-Hélène Dufour
Départment d'ethnologie
Université de Provence (Aix-Marseille l),
    general 29 avenue Robert Schuman
13621 Aix-en-Provence CEDEX 1
France

Katia Frangoudes
OÏKOS
91, rue de Saint-Brieuc
F-35000 Rennes
France

Mark Wise
Department of Geographical Sciences
University of Plymouth
Drake Circus
Plymouth
Devon PL4 8AA
UK

Serge Collet
Alte Rabenstrasse 8
20148 Hamburg
Germany

Örn Jónsson
Fisheries Research Institute
University of Iceland
Dunhaga 5
107-Reykjavík
Iceland

Gíslí Pálsson
Faculty of Social Science
University of Iceland
101-Reykjavík
Iceland

Agnar Helgason
Faculty of Social Science
University of Iceland
101-Reykjavík
Iceland

Ellen Hoefnagel
LEI-DLO
Agricultural Economics Research Institute
Burgemeester Patijnlaan 19
Postbus 29703
NL-2502 LS The Hague
The Netherlands

Bjørn Hersoug
Norwegian College of Fisheries Science
University of Tromsø
N-9037 Tromsø
Norway

Åge Mariussen
Nordland Research Institute
N-8002, Bodø
Norway

Audun Sandberg
Nordland College
N-8002, Bodø
Norway

Arne Kalland
Centre for Development and the
    Environment
Universitetet i Oslo
PO Box 1116 Blindern
N-0317 Oslo
Norway

Juan-Luis Alegret
Grup d'Estudis Socials de la Pesca
    Maritima
Departament de Geografia
Historia i Art
Universitat de Girona
Pl. Ferrater Mora, 1
E-17071 Girona
Spain

Kevin Crean
International Fisheries Institute
The University of Hull
Hull HU6 7RX
UK

David Symes
School of Geography and Earth
    Resources
The University of Hull
Hull HU6 7RX
UK

Introduction

Chapter 1
Fishing in Troubled Waters

DAVID SYMES

Introduction

Marine biologists, administrators and fishermen have lived with the knowledge of overfishing for more than a hundred years, since the first scientific evidence of declining catches per unit of effort for cod, plaice and herring in the southern North Sea came to light in the early 1890s. Less than a decade earlier one of the most eminent of Victorian scientists, T.H. Huxley, was extolling the inexhaustible resources of the oceans, urging fishermen to fish when and where they like, without let or hindrance for the resources of the sea were so vast. But the early warning signs were ignored. A contagious diffusion of overfishing spread northwards and westwards to embrace almost all of the key stocks of demersal foodfish throughout the North Atlantic by the late 1930s. The pattern of declining stock abundance was relieved temporarily by periods of recovery during and shortly after the two world wars, when the freedom of the high seas was largely constrained by risks of naval engagement – a circumstance which provided further circumstantial evidence in support of the view that increasing fishing effort was a principal contributory factor to stock depletion.

Today the condition of overfishing is endemic throughout virtually all of the world's oceans. The Food and Agriculture Organisation estimates that some 70% of the world's fish stocks are now overfished (FAO, 1995). The Worldwatch Institute (Weber, 1994) has catalogued the disastrous consequences of over-exploitation of marine fisheries around the world. Globally, the marine catch has stagnated and we appear to be reaching, or have already exceeded, the limits of the sustainable harvest of the oceans as it is commercially defined at present. 'All of the world's major fishing grounds are at or beyond their limits, and many have suffered serious declines' (Weber, 1994). Of all the world's major fishing regions, only the Indian Ocean continues to yield an increasing harvest. Many regions which underwent rapid development in relatively recent times have already begun to record sharp reversals in their fortunes; but some of the most serious declines in catch levels are in areas of long established commercial fishing. The north east Atlantic, one of the most prolific fisheries and the first world region to exhibit signs of overfishing in the modern era, has recorded one of the longest periods of sustained decline, from 13.2 million tonnes in 1976 to 11.1 million tonnes in 1992, a fall of 16%. Rates of decline have been even steeper in the north west Atlantic – 42% since 1973 – and in the Mediterranean – 25% since 1988.

The situation is even more disturbing when account is taken of the decline in individual commercial species: several of the world's most important foodfish have experienced severe cut-backs in their harvest yields. These include both the highly unstable 'r-strategists' – fish like anchovy, herring and pilchard with a short though prolific reproductive life span – and the more stable 'k-strategists' (Atlantic cod, haddock, hake, redfish and Alaska pollack). The decline in these traditional foodfish species implies that the overall expansion and more recent stagnation in the global catch has been sustained in part by the introduction of new and often much lower value species, many of which are consumed not directly as foodfish but as 'industrial' species destined for reduction to fish meal and oils.

But, as Weber (1994) graphically reminds us, overfishing is not the only form of mismanagement of the oceans. Degradation of the marine environment and inter-ference with the ecosystems through pollution, resulting from both deliberate and unconscious use of the oceans for waste disposal, together with habitat destruction arising from development in the coastal zone, are increasingly important factors in the decline of the oceans' productivity. Reduced yields and the contamination of fish, especially shellfish, by toxic chemicals and health threatening pathogens are burgeoning concerns throughout large areas of the oceans, from Chesapeake Bay to the China Seas.

The shadow of Huxley's advice still haunts the industry. Despite all the evidence from more than a century of scientific investigation, the fishing industry appears to regard the resources not simply as renewable but also as inexhaustible. Fishermen still prefer to live by the discredited myth, in studied ignorance of the scientific facts. Why is it that, more than one hundred years on from the first scientific evidence of overfishing, we seem no closer to solving the problem? It is hardly likely that the facts themselves are wrong. Evidence of overfishing and its implications for the sustainability of fish stocks continues to accumulate. Circumstantial evidence of abundant catches may, from time to time, appear to contradict the scientists' more pessimistic estimates of stock depletion and this may be sufficient to undermine the industry's confidence in the accuracy of the scientific findings. Even among scientists there is a growing wariness about the reliability of their data and the use to which they are put in stock estimation. More fundamentally, 'chaos theory' may be seen to challenge the very premise upon which science-led fisheries management is based. But the problem goes much deeper, for it is the underlying interpretation of why overfishing persists that is at fault. More particularly, the evolving systems of fisheries fail to take sufficient account of the behavioural characteristics of fishermen which are embedded in particular economic, social and cultural contexts.

Clichéd interpretations of the problem, which attempt to simplify what is a funda-mentally complex situation, have hindered rather than helped the search for under-standing. 'Too many boats chasing too few fish' is a classic example. True only at a very basic level, the statement reduces the problem to one of numbers and so makes it apparently amenable to simple forms of regulatory solution. 'Too many boats' disguises the truth about harvesting capacity and the greater importance of modern fishing technology so that crude vessel numbers and simple measures of aggregate size (ton-nage, engine capacity) and even the more sophisticated formulae combining length, breadth and engine size are no longer sufficient to describe the harvesting potential of the fishing fleet. The cliché also oversimplifies – to the point of concealment – the complexities of environment, ecology and population dynamics which govern the size and behaviour of the fish stocks. Not that fisheries policy has been constructed around such naïvely simple conceptualizations of the problem. But, by assuming that there is a basic numerical equation underpinning the relationship between fishing activity and fish stocks, fisheries policy has failed to recognize the full significance of those factors which lie outside the realms of biological science and the need to create something more than limitations on effort or output.

By ignoring both the broader environmental and institutional issues, fisheries policies and management systems have been drawn into inadequate and inappropriate 'solutions' which tend to treat only the symptoms and not the underlying problem. The endemic condition of crisis in the world's fisheries generates recurring symptoms of increasing effort and declining stocks, to be treated with familiar therapies in ever increasing strength but which leave the basic condition unaltered. The need, therefore, is for a new diagnosis of the fisheries crisis from a perspective which looks beyond the superficial facts to uncover the underlying socio-cultural conditions.

The aim of the present collection of papers is to re-examine the evidence from a social science perspective and to reformulate the agendas for research and political initiatives in the field of fisheries management. The emphasis is almost exclusively on the situation in the north east Atlantic region and the adjoining, semi-enclosed Mediterranean Sea. The authors are less concerned with arguing the case for alternative systems of regulation than with the institutional frameworks within which such new policies will need to be formulated. The remainder of this chapter will attempt to map out some of the underlying trends and examine elements of the current crisis in fisheries as evidenced by rising levels of 'civil disobedience' in the fishing industry, before exploring the trends within fisheries management in more detail. The chapter ends by outlining the structure of the following sections and indicating some of the main lines of argument contained in the book.

A crisis of science

Biological principles have dominated the basic concepts of fisheries management for close on a century. Beginning with the pioneering work of Petersen (1894) and gradually refined by Russell (1931) and Graham (1935), the concept of biological modelling of fish stocks was finally formalized in Beverton & Holt's treatise (1957), regarded as a crowning achievement in the scientific theory of fisheries management. Built on assumptions that most fish stocks are inherently stable, behave predictably under moderate levels of exploitation and tend towards an equilibrium state, fisheries management became, in principle, a relatively straightforward scientific exercise. For a given equilibrium, it was simply necessary to calculate the proportion of the adult stock that could be extracted through fishing without endangering its sustainability – in other words, the total allowable catch (TAC). The concept of maximum sustainable yield (MSY) thus became a key reference point for fisheries management. This was subsequently complemented by the work of Scott Gordon (1954) in defining the alternative management goal of maximum economic yield (MEY) to form the Gordon–Schaefer bio-economic model that has governed the basic parameters of fisheries management for almost 40 years. Its scientific legitimization, uncomplicated form and ease of application and its 'guarantee' of success in ensuring the maintenance of stock levels were sufficient to commend it to policy makers – both bureaucrats, many of whom were originally trained as marine biologists, and politicians alike.

The International Council for the Exploration of the Seas (ICES), the world's oldest intergovernmental organization for marine science, had been founded in 1902 in the wake of early concerns for the effects of overfishing. It still remains the principal source of scientific advice on stock management in the north east Atlantic region and is closely associated today with the concept of TACs. But the presentation of its advice has changed. From 1953 to 1977, the development of quantitative assessment theory, based essentially on the formulations of Schaefer, Beverton and Holt, tended towards the setting of specific catch recommendations for individual species, using biological reference points (F_{max}). Thus, unwittingly and unwillingly, the scientists began to assume the responsibility for framing management objectives for the industry. Anxious to avoid this blurring of the roles of objective scientific advice and policy making, after 1977 ICES shifted their position to one in which a range of options for TACs were set within a framework of safe biological limits, which now rests on the concept of minimum levels of sustainable spawning biomass (SSB) and incorporates an element of risk analysis.

Recently, however, fundamental weaknesses in the bio-economic model as a basis for fisheries management have been exposed. First and foremost, the theory takes little account of the tendency towards instability within the oceanic environment; it also clearly oversimplifies the behavioural characteristics of different fish stocks (Caddy & Gulland, 1985) and ignores the complex species interactions within the marine eco-system through its insistence to date on reference to single species; and it fails to recognize the disruptive effects arising from the complex dynamics of scarce resources, technological development and human behaviour.

There is increasing evidence of the unforeseen consequences of environmental per-turbation, where resource depletion may be triggered by natural events but compounded by the effects of overfishing (Glantz, 1992). One of the most dramatic examples was the collapse of the Peruvian anchovy fishery in the early 1970s when catches slumped from 13.1 million tonnes in 1970 to less than 2 million tonnes only four years later. Here it seems that the inherently unstable nature of the species, exacerbated by the effects of very high fishing intensities, reacted violently to elevated sea surface temperatures and reduced primary production associated with El Niño (Caviedes & Fik, 1992). Similarly, the sharp downturn in catches of Barents Sea cod in the second half of the 1980s appears to reflect a combination of circumstances: slight changes to the oceanic environment causing a southward extension of the winter ice front affecting the reproduction and recruitment of capelin, the principal prey species of the cod, together with very heavy fishing pressures on both capelin and cod. Both these episodes have reasonably happy outcomes with the eventual recovery of stocks and harvest yields.

By contrast, questions surrounding the underlying causes and the long-term effects of the present crisis affecting the cod fisheries in the north west Atlantic are bound to cast doubt on the validity of an equilibrium theory. No complete explanation of the collapse of the northern cod stock in NAFO areas 2J3KL at the end of the 1980s is so far forthcoming. Several possible causes have been identified: the most plausible being a cold water anomaly affecting either the spawning, recruitment, growth or migration of the cod, already under severe pressure from fishing. The significance of this case, as compared to the Barents Sea cod, is the absence to date of any real signs of recovery. Pessimists talk gloomily of the prospect that the cod may never return in commercially viable numbers to a fishery which not very long since was regarded as one of the most prolific and reliable in the world. The episode also provides an example of the contra-dictory evidence of scientists and fishermen – but, on this occasion, it was the scientists who wrongly predicted increasing stock abundance while local fishermen were warning of a profound disturbance to the cod stock as evidenced by poor catches and retarded growth weights among marine fish (Finlayson, 1994).

These cautionary tales would seem to indicate that little can be taken for granted in the management of fish stocks and especially those that are already weakened by over-exploitation. Similar population crashes are likely to increase in future as the result of climatic changes induced by global warming. However, Glantz (1992) points out that

> 'climatic change will have an uneven impact on the marine environment ... some commercially exploited populations will decrease and collapse in response to environmental changes, while others might expand and prosper in response to those same changes.'

Fisheries management would thus seem to be moving towards a period of heightened uncertainty.

But an even more formidable challenge may await fisheries scientists and managers in the form of 'chaos theory' (Smith, 1990; Wilson & Kleban, 1992). The well established Newtonian interpretation of natural systems as 'in periodic order', which is used to validate the basic conditions of the bio-economic paradigm, is now confronted by a view which holds Nature to be 'non-random but unpredictable'. In such circumstances, fish populations will not exhibit an equilibrium tendency but will vary unpredictably within certain limits. Serious doubt is thus cast upon the accuracy and value of annual estimates of stock abundance based on assumptions of linear relationships and predictability and also upon management systems which rely on the annual fine tuning of sustainable fishing effort by means of TACs and quota adjustments. So far, chaos theory is unable to offer much by way of constructive advice on how the regulatory systems might be restructured, beyond Wilson and Kleban's 'ecologically adapted management' derived from knowledge of the more stable, long-term relationships within local ecological systems. Both Smith (1990) and Wilson & Kleban (1992) agree that chaos theory comes much closer to the fisherman's own experiential perceptions of Nature as unpredictable. Management systems born of chaos theory are likely to involve less regulation and allow greater flexibility of response, much more in keeping with traditional fishing strategies.

Fisheries management is 'management under conditions of extreme uncertainty' and it is difficult to escape the conclusion that a system based on the application of predictive science to single species is likely to get it wrong almost as often as it gets it right. Although challenged in theory and discredited in practice, bio-economic theory still remains the best available (McGoodwin, 1990). In terms of its scientific basis, fisheries management faces an uncomfortable dilemma. As Holm (1995a) points out, 'where the simplifications of the single species model destroy its predictive capabilities, the realism of multi-species models create unmanageable complexities'.

Modernization and the crisis of institutions

What the current scientific debate signals is that the fishing industry faces a crisis of management, rather than a crisis of resources. Interwoven into this is a number of threads relating to scientific knowledge – its content, interpretation and use – and the institutional systems through which fisheries policy is formulated, disseminated and implemented. Just as the systems of scientific knowledge, which underpin modern fisheries management, are coming under increasingly critical interrogation, so too are the institutional arrangements which have developed over the past 40 years or so largely to ensure the delivery of science-based, technocratic policy.

Underlying a condition of what might be called 'institutional turbulence', we can recognize perhaps three broad tendencies of modernization which, though certainly not unique to fisheries – they are, for example, strongly evident in the problems facing modern farming systems – have certainly had a severe impact on the industry. These tendencies are:

- the penetration of capital through industrial modes of exploitation, formerly identified mainly with the offshore sector and especially with distant water fishing nations like the UK, but now increasingly evident in the inshore fisheries and in aquaculture;
- the appropriation of responsibility for fisheries management from local, industry-based institutions and its relocation in the corridors of bureaucratic power in central government departments and supranational authorities like the European Commission;
- the globalization of the food system as a result of which local fisheries no longer enjoy a monopoly in local or national markets but are increasingly engaged in intensive competition with sources of supply across the world.

Each of these tendencies progressively marginalizes the fishermen and their organizations as economic and social actors throughout the range of activities from harvesting to marketing of the fish.

From these three broad tendencies we can define a cluster of more specific, but still interrelated, crises which confront the industry. The first is a *production crisis*, initially identified in terms of the contagious spread of overfishing to almost all parts of the oceans. Whilst overfishing is clearly of direct concern to the fishing industry, the fishermen's own perception of the production crisis lies in the increasing constraints placed on their traditional freedom of action through regulations concerning TACs, quotas, gears, closed areas, etc. Each of these regulations is seen as a deliberate constraint upon the return on investment in new technology and upon the professional skills and flexible strategies of the fisherman (Vestergaard, 1994). In short, the problem of scarce resources is compounded by constraints imposed on the efficiency of the enterprise.

Linked to this – and in some ways a part of the production crisis – is the growing *crisis of property rights* and the creeping threats to the fundamental concepts of common use rights and open access, which have largely structured traditional perceptions of the fishery and its management and underpinned the value systems of fishing communities. Freedom of fishing on the high seas has been whittled away by the extensions of sovereignty up to the 200 mile limits and, within these newly prescribed national fishing territories, by the adoption of restrictive licensing systems and 'closed areas'. The threat of privatization of the commons through the introduction of individual transferrable quotas (ITQs) – favoured by economists and many administrators as a logical extension of restrictive licensing and quota management systems – poses perhaps the greatest threat to established perceptions of marine resources.

The *crisis of the markets* describes a small constellation of problems resulting from the globalization of the food system and the transfer of power in the market place from the producer to the secondary processors and, more especially, the multiple retailers who today control a very high proportion of the end sales of fish in Europe. In the face of changes in the balance of power and as a result of the growing deregulation of agricultural commodity prices, first-hand sale prices for fish face strong downward pressure (see Chapter 16) with the result that reduced levels of catch are no longer compensated by rising quayside prices. At the same time, the growth of convenience foods and the commanding position of the supermarkets have presented an opportunity for the substitution of popular local species by less familiar ones, as with the substitution of Atlantic cod and haddock by Alaskan pollack and New Zealand hoki. Although consumer prices for fish products remain high, in comparison with alternative foods, the price paid for the raw material has been depressed and the stability of local markets has been disrupted by sudden surges in imported supplies of cheaper frozen fish. Thus, the local inshore fishermen, once regarded as the unrivalled source of high quality prime fish for the retail market, have been dispossessed of this niche role by the changing structure of the consumer market.

Key elements of all three situations are drawn together in a fourth *crisis of the institutions* (or a crisis of management). Traditional forms of social organization for the management of fisheries have been abrogated. Customary systems of management have been replaced by centralized bureaucratic policy making which, in most but not all instances, excludes the fishermen's organizations from an active influential role and has singularly failed to carry conviction and win support among the resource users. The contribution of generationally transmitted knowledge and experiential understanding has been replaced by experimental research, sample surveys and the linear programming of the results to estimate future fish stocks. In policy terms, this technocratic approach to the understanding of the resource has led to the substitution of flexible strategies, developed in the context of particular local fisheries, by rigid regulatory frameworks applied over much larger territories.

All four circumstances outlined above may be said to precipitate a further *crisis of confidence* which pervades the industry and threatens the breakdown of order within the fishery. Principally, it is the fisherman – deprived of his customary rights, denied the status and prestige due to his professional knowledge, skills and experience, and sidelined in the management process – who has lost confidence in his own identity and value systems, as well as in the institutions of fisheries management and their outcomes. McGoodwin (1990) suggests that, because of the nature of their work, fishermen may sometimes find themselves estranged from the web of social life in their own communities and thus at a disadvantage in sustaining their own representative organizations and building wider networks of political alliance beyond the local community. Evidence from this volume (see, for example, Chapters 9, 10 and 11) would suggest that Europe's fishermen do not lack the skills of effective political organization at the local level, though their influence on the wider political stage may be very much weaker. There are signs that the policy makers themselves and their advisers, profoundly disturbed by the unremitting evidence of the failure of existing policy measures, are also losing confidence in the present system and are prepared to seek new forms of institutional arrangement.

The near anomic condition of the fishing community and its alienation from the management system has sparked off direct action and civil disobedience both on land and at sea. Witness, for example, the actions of Breton fishermen in protest at the collapse of quayside prices (see Chapter 13) or the blockade of ports by Sicilian and Calabrian swordfishermen over the imposition of 'external' regulations (see Chapter 10). Recourse to the legal process of challenging policy decisions in the courts – as occurred recently in England over the proposed introduction of days at sea restrictions – is expensive and far less common than the increasingly frequent, almost epidemic, acts of civil disobedience – the classic symptoms of institutional failure.

The breakdown of law and order, together with the criminalization of fishermen in the portrayal of events by the media, takes a number of different forms. It occurs in the persistent disregard for rules as in the infringement of gear regulations and minimum landing sizes and the contravention of fishing limits; it may involve massive breaches of quota regulations in the form of blackfish landings. Ironically what may be construed as delinquent behaviour may be legitimated by the regulations themselves, as with discards of fish below the legal landing size or above the quota entitlement of the vessel.

Circumventions of international high seas agreements are also commonplace, often involving the reflagging of vessels under a non-signatory state so as to put themselves outside the law. The existence of 'loop holes' in the political geography of the high seas through the underlapping of exclusive economic zone boundaries has helped to provide safe havens for vessels engaged in poaching activities within national waters.

Significantly, it is not only individual fishermen who flout international law but, on occasions, the nation state itself. The recent, highly publicized dispute between Canada and Spain in the north west Atlantic is a case in point where Canada briefly tried to usurp rights of control beyond the 200-mile limit in order to counteract allegedly illegal high seas fishing by Spanish vessels on the 'Nose and Tail' of the Grand Banks. In the same geographical context, the EU's refusal to accept NAFO quotas in the north west Atlantic – where the EU is itself a contracting party – may be cited as an institutionalized disregard for international agreements.

Industrial fisheries and customary management systems

Confrontation between the more egalitarian, customary modes of local management and the processes and competitive ethic of capitalism are a key to the understanding of the fisheries crises and the consequent breakdown of law and order. The intervention of industrial capital has trespassed on traditional territories, transgressed basic rules and replaced a concept of collective sustainability by competitive exploitation. This is a recurring theme in the papers collected together in the present volume. We must be careful, however, not to overstate the efficacy of local management systems; they were much less widespread than is commonly believed and by no means all were successful. Where they did succeed in achieving a sustainable fishery, the local management systems were usually introduced through collective agreements, embedded in customary social regulation and maintained through respect for potentially severe sanctions. Codes of behaviour normally applied only in inshore waters and were based on the recognition of customary territorial use rights, rather than formal property rights, together with the principle of open access for all members of the territorial community. Complex rules governing the time and duration of access and defining the permitted gears were enforced by strict sanctions which could ultimately disbar the transgressor from participation in the fishery. Such regulations were often finely tuned to take account of increasing or decreasing numbers of participants and/or improved catching capacity.

The survival of local systems of management – or, in many more instances, the persistence of small-scale fishing in conditions of unmanaged exploitation – was challenged and eventually overwhelmed by the tendencies of modernization and especially the processes of capitalist penetration, accumulation and concentration affecting the means of production, technological investment and the creation of global markets. Capital has sought to delay the economic consequences of declining resources by technological innovation and 'capital stuffing'. This was especially evident in the growth of distant water fishing interests among the more industrialized countries bordering the north east Atlantic. Although industrialization came much later than in many other sectors of the economy, countries like the UK and Germany came to rely heavily upon supplies of both wet and frozen fish landed by large, modern, distant-water trawlers in company rather than individual ownership. In 1968, more than 35% of the supplies of demersal foodfish for the UK markets were landed by vessels working the distant waters. The smaller-scale, family-owned enterprises were gradually drawn into competitive, capitalist relations and their fishing gears were subject to increasing technological sophistication, enhancing their catching capacity and thereby threatening the integrity of the local regulatory system from within.

A period of unequal and uneasy co-existence for the small-scale 'artisanal' and the capital intensive 'industrial' sectors lasted only while marine resources remained relatively abundant and markets for their catches were strongly differentiated. Conflict increased with the onset of both resource and market crises and the attempted imposition of universal, technocratic solutions through the intervention of the state in the regulation of production and bureaucratic systems of implementation. Science was used to legitimate a system of regulation (TACs and quotas) that supports the *status quo* (Gulland, 1987), shields the policy makers from having to confront awkward questions relating to the basic objectives of fisheries management and coincidentally protects the interests of capital. Certainly, global concepts of MSY and MEY were more readily appreciated by corporate interests than by small-scale fishermen engaged in a daily struggle for survival. TACs and quotas suited those with enhanced harvesting capacities by both forcing up the price of scarce resources and giving the advantage to those with the gear and vessel technology to make a quick kill. More recently, the introduction of ITQs clearly favours those with capital, collateral and easier access to the money markets.

Moves to protect the interests of coastal states through the extension of fishing limits to 200 miles did not automatically enhance the status of the inshore sector. Intended to relieve the pressure on stocks from foreign-based, distant-water fleets, the creation of EEZs led to a displacement of fishing effort. While some vessels were either scrapped or sold out of the fleet, others were diverted to the redefined 'high seas' beyond the 200 mile limit, often targeting straddling stocks which lay partly inside and partly outside the EEZ, as in the case of the Nose and Tail of the Grand Banks. In some cases, powerful distant-water vessels were redirected onto stocks in home waters which had previously been the virtual preserve of the inshore fleets, thus intensifying sectoral conflict. Decline of distant-water fisheries in many cases stimulated a renewal of capital investment in middle-water and inshore vessels, helping to accelerate the diffusion of technology. Nominal beneficiaries of the new political geography of the oceans – countries like Iceland and the Faroes, which had built their fishing industries on the basis of small-scale inshore vessels – witnessed an explosive growth of their domestic fishing capacity, often financed through the processing industries, and a steady erosion of the resource base.

Not all modern management systems exclude fishermen from policy making. The 'negotiation economies' of Denmark and particularly Norway have provided opportunities for the strong representation of the resource users' views in the consultation procedures which precede policy decisions – somewhat in contrast to the centrally managed policy formulation in the UK where no formal consultation procedures exist. But it is clear that, in all cases, policy making systems increasingly favour the interests of capital – both land-based and sea-going – at the expense of the small-scale producers. Even in Norway, the strength of the fishermen's representation in the decision-making system has been squeezed by business interests in reforms since 1977 (Holm, 1995b). The contribution of small producers to overall output is frequently underestimated and certainly undervalued. Part-time fishermen are particularly disadvantaged: they may be excluded from licensing systems and decommissioning schemes and fail to qualify for specific forms of socio-economic assistance. Yet these enterprises are rarely, if ever, directly responsible for the overfishing of stocks. Within the north east Atlantic there is little evidence – except in Norway – that the state has intervened on the part of particular regions or social groups to offset the concentration of capital by preferential quota allocations or direct subsidies to the small boat sector.

One may indeed marvel at the survival of small scale enterprises in conditions which are increasingly hostile. Growing regulation of the fishery usually implies that men must work harder for less return, with increased risks of financial failure, unemployment and poverty – scant reward for involvement in an industry which is among the most physically demanding and hazardous, with unsociable conditions and patterns of employment often involving lengthy separation from their families. There are some similarities with the survival of 'small commodity producers' in agriculture, though it would be dangerous to push the comparison too far. Although we can draw some important parallels – dependence on family-based structures, importance of high instrumental values and victimization at the hands of the commoditization process – equally, there are significant differences. Perhaps the most fundamental concerns property. De Haan (1994) has sought to explain the persistence of small family farms in the Netherlands in terms of the cultural and symbolic value of land. For fisherman symbolic value is much less tangible. Except where ITQs have been introduced, private property in the harvesting sector is usually confined to the vessel and its gears.

For small scale fishermen one of their most valued assets is their intellectual capital, consisting of detailed knowledge of the local fishing grounds and the behaviour of the fish stocks (see Chapter 11), gained through intergenerational transfers and experience. It is this specific knowledge that helps small-scale fishermen survive in competition with larger, technologically sophisticated vessels which often lack such knowledge (McGoodwin, 1990). The absence of property and tradable assets also restricts the options in terms of survival strategies. Whereas in farming, diversification of land use to non-agricultural ends has become a central theme of policies for rural restructuring, such options are not widely available to fishermen. In their case the choice is simpler: expansion through investment in improved technology in order to achieve short-run advantages in the race to fish; retrenchment implying increasing levels of self-exploitation and falling profit margins; or disengagement from the industry. Only the last two are assured in the present conditions.

Outlining the argument

It should be increasingly clear to all – fishermen, political decision makers and academic commentators alike – that the *status quo* is no longer an option for the future of fisheries management. Current regulatory packages and the institutional systems that support them have been largely discredited by the widespread evidence of policy failure throughout the world. In the view of many social scientists, the search for solutions must focus initially on institutional restructuring (see, for example, Jentoft, 1989; McGood-win, 1990; Townsend, 1995). That is the principal theme running throughout the remainder of this book. Although there is no intention to frame specific recommen-dations, certain broad conclusions will become apparent and these will be re-examined in the final chapter.

The intervening chapters are divided among five sections, each reflecting a key area of concern for the social sciences. The first refers to perhaps the most difficult of all the issues confronting the policy makers – the reformulation of the objectives for fisheries management and, in particular, the need to incorporate a social dimension in fisheries policy. Hitherto, social issues have been treated as 'externalities' of the bio-economic model and thus left to other policy areas to address. In Chapter 2 Hersoug reviews the roles of biology, economics and the social sciences through a 'dominant theory' approach, arguing that the social scientist must challenge the simple assumptions that fishermen's goals revolve around income maximization and retaining rights of access. The incompatibility of biological, economic and social goals means that choices will have to be made and compromises struck, but without social objectives to guide the distributional effects, regulatory systems will continue to be jeopardized by non-compliance. According to Mariussen (Chapter 3), actors in the industry seek to embed risks in certain forms of social organization, in co-operative action and in specific reg-ulations where social objectives have been defined at policy level. Using examples from both manufacturing industry and fisheries in Norway, he illustrates ways in which compliance with potentially adverse policies can be brokered through formal negotia-tions between policy makers and resource users. In Chapter 4 Sandberg's critique of the EU's Common Fisheries Policy (CFP) focuses upon the concept of 'commonalty': for whom are the fishery resources a common property – the EU, the member state or the region? He argues that fisheries management is best handled at a local level where the need to protect the social infrastructure is most readily recognized.

The issue of property rights is addressed in more detail in Part II. According to many the persistence of common-use rights remains a major obstacle to the development of an effective management regime: it encourages overcapitalization of the harvesting sector in the attempt to capture short-run advantages over competing vessels in the race to fish which, in turn, subverts attempts to manage the fishery through output controls. But alternative property rights systems also have their disadvantages. Iceland is the only North Atlantic state to have introduced a comprehensive system of ITQs and Pálsson & Helgason in their analysis (Chapter 5) find the economic argument in favour of ITQs not proven. Although ITQs are expected to lead to increased efficiency through a reduction in surplus capacity and a concentration of resources in the more efficient enterprises, there is evidence of inefficiency of resource use as a result of companies hoarding quotas which they cannot exploit themselves. In analysing the limited systems of ITQs for sole and plaice in the Netherlands, Hoefnagel (Chapter 6) develops the analogy with the use of 'pollution permits' whereby firms may discharge polluted waste water at levels sti-pulated in the discharge licence. As with ITQs, the licences acquire economic value in the transfer of licences from one firm to another. The key question in both instances is whether the market can reasonably replace government regulation in the effective management of renewable resources. In marked contrast, Kalland (Chapter 7) offers an insight into the system of territorial use rights developed in Japan on the basis of defined rights of access to specific areas of sea available to particular permit holders. These ancient rights survived the collapse of feudalism and are today encoded in formal co-operatives which take precedence over individual and private firms in the management of inshore waters.

Among many social scientists working on fisheries related issues, there is a growing concern that the marginalization of local social institutions for the management of inshore fisheries contributes significantly to the growing demoralization and lack of commitment to broader regulatory systems on the part of the fishermen. This point of view is propounded in Part III with the use of examples from within the EU. Vestergaard's paper (Chapter 8) provides a general introduction to problems posed for traditional socio-cultural systems by the imposition of modern, bureaucratic systems of fisheries management. In particular, the concept of quotas which assume predictability of catches contradicts the experience of fishermen and denies them the use of established means of coping with a fluctuating resource, building up resentment and resistance to the new regulatory systems. The other three papers illustrate the threats to local management institutions and systems in greater detail. In Chapter 9 Alegret examines the attempts to replace the unique social institution of the *cofradía* in Spain by producers' organizations. In Chapter 10 Collet takes the example of the swordfish in the Straits of Messina to argue that global regulations, advocated by the UN and endorsed by the EU, lack the sensitivity of local management systems. In Chapter 11 Dufour describes the role of *prud'homies* in Mediterranean France, indicating how these institutions, very well adapted to the diversity and fragility of the local marine ecology, are now threatened by the development of new institutional frameworks under the CFP.

Issues relating to fisheries policy extend far beyond the management of fish stocks and controls on fishing effort. Fishing communities have evolved particular structures and patterns of social interaction to accommodate the demands of fishing activity, while the development of many coastal regions is closely bound up with the fortunes of the fishing industry. Both situations are to some extent threatened by the uneven distributional effects of the regulatory mechanism. In Part IV, the impacts of the CFP are examined in a community and regional context. In Chapter 12 Frangoudes describes how modernization of the industry threatens to undermine the collective ethic in Greek fisheries and how traditional forms of negotiation between the central state and the local community, often conducted through informal channels, are being altered by the appearance of a new actor in the shape of the EU. In their analysis of the crisis in the Breton fishing industry (Chapter 13), Delbos & Prémel express a similar concern that bureaucratic forms of regulation of the fisheries and the globalization of the market are undermining the traditional structures and social networks within the Breton fishing village and causing the marginalization of the fisherman as a social actor within his own community. In Chapter 14 Wise transfers attention to the fisheries dependent regions and asks whether regional management strategies for fisheries can work in an industry characterized by conflicting interests and intra-regional tensions. He acknowledges that political alliances such as the Atlantic Arc do have a role to play in trying to prevent the further economic marginalization of Europe's coastal peripheries. Mørkøre's paper (Chapter 15) introduces the particular problems of a small, remote semi-autonomous region – the Faroes – so far not subordinated to the CFP but where fisheries politics have been dominated by divisive issues of membership and closer relations with the EU for some 20 years.

Part V deals with a wholly different set of issues which impinge upon the status of the harvesting sector, namely the globalization of the markets for fish and fish products and the impacts on the fishing industries of the North Atlantic. As the two papers reveal, the creation of a global market has important consequences for the traditional supply countries in the North Atlantic, challenging their dominant position in terms of trade with the major European markets. A detailed examination of the changing pattern of relationships within the food system and the shift in economic power from the producer to the retail sector of the food chain is provided by Friis in Chapter 16. Europe's fishing industry no longer holds a monopoly position in the supply of raw materials and finished products to the consumer markets and thus its fishermen are no longer key actors in supply and price determination. Jónsson's paper (Chapter 17) continues the analysis of changing producer/consumer relations, arguing that the interdependency of the North Atlantic region has been greatly altered by events on a global scale. He draws attention to the increasing polarization of fish products into standardized, industrial commodities, on the one hand, and specific luxury foods, on the other.

The concluding section of the volume summarizes the main findings drawn from the contributors' chapters and from this material an agenda for further research is set out with particular emphasis on co-management, property rights and high seas issues.

References

Beverton, R. & Holt, S. (1957) *On the Dynamics of Exploited Fish Populations.* Fisheries Investigations Series 2 (19), Fisheries and Food Department of the Ministry of Agriculture, London.

Caddy, J.F. & Gulland, J.A. (1985) Historical patterns of fish stocks. *Marine Policy,* **7** (4), 267–78.

Caviedes, C.N. & Fik, T.J. (1992) The Peru–Chile eastern Pacific fisheries and climatic oscillation. In *Climate Variability, Climate Change and Fisheries* (ed. M.H. Glantz), pp. 355–75. Cambridge University Press, Cambridge.

De Haan, H. (1994) *In the Shadow of the Tree: Kinship, Property and Inheritance Among Farm Families.* Het Spinhuis, Amsterdam.

Finlayson, A.C. (1994) *Fishing for Truth: A Sociological Analysis of Northern Cod Assessments from 1977–1990.* St. John's, Institute of Social and Economic Research, Newfoundland.

Food and Agriculture Organisation (1995) *The State of the World's Fisheries and Aquaculture.* FAO, Rome.

Glantz, M.H. (ed) (1992) *Climate Variability, Climate Change and Fisheries.* Cambridge University Press, Cambridge.

Gordon, H.S. (1954) The economic theory of a common property resource: the fishery. *Journal of Political Economy,* **62**, 124–42.

Graham, M. (1935) Modern theory of exploiting a fishery and application to North Sea trawling. *Journal du Conseil International pour l'Exploration de la Mer,* **10**, 264–74.

Gulland, J.A. (1987) The management of North Sea fisheries: looking towards the 21st century. *Marine Policy,* **11**, 259–71.

Holm, P. (1995a) Fisheries management and the domestication of nature. Paper presented to the Fifth Annual Common Property Conference: Reinventing the Commons. Bodø, Norway, 24–28 May, 1995.

Holm, P. (1995b) Productionism, resource management and co-management: phases in the development of Norwegian fisheries. Paper presented to the XVI European congress on Rural Sociology. Prague, 31 July–4 August, 1995.

Jentoft, S. (1989) Fisheries co-management: delegating responsibility to fishermen's organisations. *Marine Policy*, **13** (2), 137–54.

McGoodwin, J.R. (1990) *Crisis in the World's Fisheries*. Stanford University Press, Stanford.

Petersen, C. (1894) On the biology of our flatfishes and on the decrease of our flatfish fisheries. *Report of the Danish Biological Station*, **4**, 1–85.

Russell, E. (1931) Some theoretical considerations on the overfishing problem. *Journal du Conseil International pour l'Exploration de la Mer*, **6**, 3–20.

Smith, M.E. (1990) Chaos in fisheries management. *Maritime Anthropological Studies*, **3** (2), 1–13.

Townsend, R.E. (1995) Fisheries self-governance: corporate or co-operative structures. *Marine Policy*, **19** (1), 39–45.

Vestergaard, T.A. (1994) Catch regulation and Danish fisheries culture. *North Atlantic Studies*, **3** (2), 25–31.

Weber, P. (1994) *Net Loss: Fish, Jobs and the Marine Environment*. Worldwatch Paper 120, Worldwatch Institute, Washington.

Wilson, J.A. & Kleban, P. (1992) Practical implications of chaos in fisheries: ecologically adapted management. *Maritime Anthropological Studies*, **5** (1), 67–75.

Part I
The Social Dimension of Fisheries Policy

Chapter 2
Social Considerations in Fisheries Planning and Management – Real Objectives or a Defence of the *Status Quo*?

BJØRN HERSOUG

Why management?

In its simplest form 'fisheries management' can be defined as 'the pursuit of certain objectives through the direct or indirect control of effective fishing effort or some of its components' (Panayotou, 1982). Even if 'scientific fisheries management' is a western invention, the general idea that fisheries have to be managed in some way or other is found all over the world, even in so-called 'unregulated fisheries'. Rather often the terms 'management' and 'development' are confused. Management is often called upon only after a fishery becomes over-exploited, while development should be applied when a fishery is 'underexploited', that is when catches are small or way below what the resource can possibly yield. The two concepts are in fact closely interrelated. Management is required long before a fishery is biologically over-exploited, while development is absolutely called for when a resource is in fact over-exploited, not least in order to channel people and economic resources to other sectors. The general idea is that management without some elements of development is impossible and vice versa. Both 'management' and 'development' are aiming at some kind of optimum or best possible use of the fishery resource.

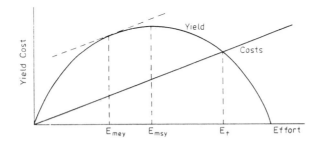

Fig. 2.1 The Gordon–Schaefer model.

How this optimum or best possible use is defined depends, of course, on the specific objectives set for a particular fishery, fishing sector or economic sector of a particular country. In general terms the objectives can be defined as biological, economic or social. If the policy objective is maximum fish production, the optimum exploitation of the fishery resource is defined as maximum sustainable yield (MSY), that is the maximum catch that can be obtained on a sustainable basis. Based on a simple Gordon–Schaefer model (see Fig. 2.1), taking one single stock into consideration, it can be shown that fishing effort should be expanded until E_{msy} where the stock is yielding at the maximum level. However, this biological objective has later been challenged by the fisheries economists, pointing to the obvious fact that fishermen most often catch in order to obtain income, not fish *per se*. Therefore income and costs have to be calculated and maximum economic yield (MEY) is reached by optimizing the difference between total income and total costs, in this case by the effort of E_{mey}. Note that most often E_{mey} is reached by considerably less effort than E_{msy}. But again the economic objective was found to be too narrow. In order to take into consideration the social effects of fisheries management, the objectives of maximum social yield (MSocY) or simply optimum yield (OY) have been introduced. Even if it is difficult, or indeed impossible, to speak of a true optimum in this latter case (depending on the different political constraints, all sorts of optima can be constructed), the general idea is that economic yield can not be considered without relation to distributional effects. Social considerations may therefore modify the pure economic solutions. When, for example, no other means of occupation is available, management may justify a more intensive rate of fishing than is arguable on purely economic grounds (Panayotou, 1982).

The beginnings of fisheries management have almost invariably been biological in nature. As pointed out by Hannesson (1993), biologists have been called upon in order to assess stocks and make recommendations as to how much should be caught. The methods have been either direct restrictions on catch or limitation on fishing activity in order to reach the same goal. Increasing pressure on the fish stocks and reduced yield have been the 'rationale' behind the biological regulations – still in effect almost all over the world. However, biological regulations do not consider the economic aspects of fishing. The labour, equipment and financial resources employed in fishing could be employed otherwise, perhaps generating more income. Moreover, many of the biological regulations tended to involve even more costly fishing, making the whole fishery less profitable seen both from the point of view of the individual fisherman and from society. Therefore, fisheries economists have tended to advocate the reduction of fishing effort, in order to take the maximum catch with the least possible effort. Finally, the social scientists have challenged the simple assumptions lying behind the economic models. Open access is not necessarily the 'natural' situation in most fisheries; fishermen do not always try to maximize their income, irrespective of what other fishermen do. In general, it is difficult to optimize without a view to distributional effects and to what is considered just and socially acceptable.

The use of goals in fisheries management

Ultimately, practical fisheries management has to deal with biological, economic and social goals. These goals, as simple as they appear in theory, are in practice very difficult to operationalize. As pointed out by Hilborn & Sibert (1988): 'when we search for the optimum level of fishing effort, we are chasing a moving mountain top in foggy weather'. In addition, most fisheries are multispecies fisheries, with complicated interactions taking place among different species. This complicates the simple modelling and adds further problems to the distributional aspects. In the end, the knowledge of the different species is far from complete, which is evidently shown by the breakdown of the Peruvian anchovy fisheries. Twenty years after the incident, biologists are still arguing over what was due to natural fluctuations and what was due to overfishing or fishing effort. Secure proof can never be found as long as we are not able to construct an experiment where the environment is controlled. Therefore, some kind of adaptive management is required.

In most fisheries planning, in developed as well as developing countries, we seldom find MSY, MEY or MSocY expressed as ultimate objectives. More often we find goals like:

- increased (local) food supply (more fish for a cheaper price);
- increased export earnings;
- increased income for the fishing population;
- increased employment possibilities for the fishing population.

As shown by Bailey & Jentoft (1990) these goals tend to be incompatible. Referring to the same limited stock of fish it is hard to increase local food supply and at the same time increase exports; increased income for the fishing population may easily contradict the goal of increasing the food supply which requires lower instead of higher prices for fish. The possible contradictions are numerous and only in exceptional cases can all four goals be fulfilled at the same time. Furthermore, the fishing industry consists of many actors with different interests, each trying to realize his (or her) particular objectives. Sectoral plans, be they on a national, regional or local level, will therefore reflect a compromise. However, seldom will this compromise be made explicit. In addition, what is stated in the plan may be different from what is implemented in practice. The competing interests are normally not settled once and for all as long as implementation is pending. Consequently, fisheries managers are left in a rather difficult situation. The techniques and models they have learnt in theory are difficult to translate into practice, due to the enormous complexity of most fisheries, and the lack of adequate information.

On the other hand, the practical goals they are given by the politicians are usually very general and normally presented without explicit priorities, so that fisheries managers are left with great discretionary powers. In the end they have to opt for management goals guaranteeing sustainable use of the resources, yielding an acceptable profit both to the individual fisherman and to society, and a distribution which is considered just and fair, at least by the dominant part of the fishing industry. So general are the goals that some biologists have even proposed management without reference to specific goals.

The nature of political objectives

For many fisheries managers the outright lack of objectives has been rather frustrating, leading to a demand for more exact objectives, preferably ordered by priority. Among social scientists working within the field of organizational theory or political science, this demand is considered out of reach. Political goals are inherently broad, trying to comprise as many concerns as possible. Neither in fisheries, nor in any other important sector of the society, is it possible to obtain exact priorities on the level of development objectives. Often we find that official authorities have one set of goals, the (private) industry another and the managers a third. Compromises are worked out continuously, but not as an open democratic process. The process of arriving at a compromise is characterized by the use of power, knowledge and legitimacy. Power is translated in terms of resources (numbers/members, capital, organizational resources), knowledge may be considered a special type of resource or alternatively an asset which can be acquired by other means; while legitimacy deals with both the process of arriving at a decision and the final result.

Some managers tend to think that goals (or objectives) belong only to the planning phase and that the battle is won when a particular goal enters the management plan or the development programme. Many studies from the implementation literature give evidence to the contrary; as soon as the final (planning) decision is made, the battle of implementation starts where most participants again try to follow their own particular interests.

This is not to argue that goals are without interest in fisheries planning and management, only that goals are fluid and constantly changing, depending on the strength of the actors or groups participating. Furthermore, official goals very often give little insight into what is really going on in the fishing sector. Consequently implementation should be considered a field of its own or as a continuous process of policy change.

In order to understand the implementation process, three broad perspectives can be described. The first concentrates on implementation as a symbolic activity. The case (the project or programme) is not what matters, but rather the opportunity to present oneself. The participants are eager to create legitimacy through what is said, how it is said and to whom. The second perspective describes implementation as interest group politics. Here the formal project or programme agreement does not mark an end for the competing groups. Instead, they continue to follow their own interests, depending on the resources available and the coalitions that can be made. The implementation process can therefore be seen as a continuous negotiation process. In the third perspective implementation is a social by-product, where changes in the environment outside the control of the organizations lead to new decisions and changes in the original project or programme. The outcome is a by-product of the changes in the environment, where the principal actors have only limited control.

The three perspectives are complementary and may be used to analyse the outcome of projects or programs. The main lesson in this context is that goals mark only the beginning; the importance of goals is shown through the process of implementation.

The Norwegian experience

Even if Norway has had a separate Ministry of Fisheries since 1946 (and a Fisheries Directorate from 1906), the first fisheries plan, comprising the whole sector, appeared as late as 1977. The occasion was the establishment of the 200-mile exclusive economic zone (EEZ), giving Norway the possibility of planning the use of the resources inside the 200-mile zone. The plan centred on three main objectives:

- securing the sustainability of the fisheries resources;
- securing the present settlement pattern;
- securing employment at sea as well as on shore.

Practically everybody agreed that the industry, both in terms of vessels and production plants, was too large in relation to the fishing resources available and, furthermore, that the capacity had to be reduced. The disagreement entered into the question of who were to quit: small-scale or large-scale enterprises; enterprises in the north or in the south; enterprises based on active or passive gear? The plan ended by recommending a drastic reduction of the industrial fleet, thus favouring the 'social' objectives of the plan. However, in the implementation phase the political majority faded away and when the resource prognosis had to be altered few years later, the whole plan was soon shelved. It should also be noted that the generality of the objectives was hard to handle in practical policy. Was the settlement pattern to be maintained at the level of the local community, the municipality, the county or on a regional level? How much income could be sacrificed in order to meet such an objective?

By the time of the next long-term plan the economists had taken the lead, and 'increased profitability for the fishing industry' became a fourth major goal. Reduction of state subsidies was an important issue, but the new long-term plan followed the usual Norwegian tradition: broad-based goals without priorities.

This does not imply that social goals like 'maintenance of the existing settlement pattern' or 'secure employment' are worthless. The point is that these broad goals are used by different groups either to push for or to resist a particular policy.

When, for example, the Norwegian Government tried to introduce a restricted version of the Icelandic individual transferable quota system, this particular measure was resisted with explicit reference to the goal of maintaining the existing settlement pattern which, in the Norwegian context, means strengthening the coastal fleet, especially in the north. Even if the offshore industrial fleet has gained strength over the last 40 years, the coastal fleet is, especially compared to other European countries, still rather strong, not least due to the political clout of the coastal fishermen and their followers. The same line of argument can be used to explain the strong support of different state subsidies in the fishing industry for many years. The support was, at least in official rhetoric, a defence of the decentralized fleet and production plants.

But social goals can also be used the other way around, as when the fishing industry demanded a revised Raw Fish Act ('the fishermen's constitution'), referring to the need for better financial conditions and more secure jobs in the fishing industry.

In the recent great debate concerning the question of joining the EU, both the 'yes' and the 'no' lobbies referred to the social goals in fisheries policy, indicating that the EU's fisheries policy would either destroy or enhance those existing goals.

Again this debate demonstrates that goals in themselves are not very interesting. The decisive question is: who is using what kind of goals in what kind of policy matters? Or, to phrase the question in a more scientific manner: who is able to present their cause in a manner that is considered legitimate by the wider public? As described above power counts but legitimacy decides. In a Norwegian context this means that the industrial ocean-going fleet has the resources, the organization and, in general, the means to gain the upper hand. Yet the coastal fleet is able to keep a considerable part of the fishing resources in terms of the share of the TACs and the economic resources as well as gaining general protection by way of law blocking a total takeover by the industrial fleet.

Social considerations – a by-product?

Many fisheries economists have argued that social considerations or goals can be treated as a special case or condition within more general economic models. It is, for example, perfectly possible to anticipate a certain level of employment and then maximize the production function. Still this procedure raises the problem of maximizing or optimizing income, without considering the initial distribution. In the end, any kind of allocation – of fisheries resources or economic resources – has to be considered just or fair, at least by a majority of the fishermen or by the operators in the industry in general.

This is not least the case with the Common Fisheries Policy (CFP) of the EU. The great problems of the CFP today are the overcapacity relative to the resources and the lack of respect for fisheries rules and regulations. Both problems are closely related to the social aspects of the CFP. Reducing the capacity of the fleet by say 40% will necessarily cause large social problems, creating unemployment in marginal regions with few alternative opportunities. When such a reduction has been proposed by the Commission (DG XIV), the issue has immediately been turned back by national political representatives as well as by representatives from the industry. Preferably, nobody should be forced out of business; alternatively, costly compensation schemes have to be established. In this manner, social goals or social considerations may be interpreted as another phrase for the *status quo*.

On the other hand, EU policy has to introduce more explicit social goals in the CFP. If not, the whole system of rules and regulations will break down for lack of compliance. When the majority of fishermen feel that regulations are illegitimate, they will not stick to the rules, no matter how much control is being introduced. Holden (1994) stresses the management goal of MEY and more sophisticated control measures as the solution to the present problems. This solution is much too simple. The CFP has to introduce more explicit social goals and, if these goals are going to have any impact, they must have sufficient political backing throughout the implementation phase. With a limited resource base a more equitable distribution will necessarily imply a redistribution of quotas and funds, maybe including restrictions on the type of technology to be used in different fisheries. In the end a workable solution will have to be found, striking a compromise between what is a sustainable use of the resources, a reasonable level of efficiency for society as a whole, as well as for the individual operator, and a distribution of income which is considered just and fair, at least among the majority of participants in the industry.

References

Bailey, C. & Jentoft, S. (1990) Hard choices in fisheries development. *Marine Policy*, **14**, 333–4.

Hannesson, R. (1993) *Bio-economic Analysis of Fisheries*. Blackwell Science Ltd, Oxford.

Hilborn, R. & Sibert, J. (1988) Adaptive management of developing fisheries. *Marine Policy*, **12**, 112—21.

Holden, M. (1994) *The Common Fisheries Policy*. Blackwell Science Ltd, Oxford.

Panayotou, T. (1982) Management Concepts for Small Scale Fisheries: Economic and Social Aspects. *Fisheries Technical Paper*, **228**. FAO, Rome.

Chapter 3
Social Objectives as Social Contracts in a Turbulent Economy

ÅGE MARIUSSEN

Introduction

Institutional structures have to be organized to take account of risk emanating from turbulent conditions in the global environment (Granovetter, 1985; Giddens, 1990; 1991). The fishing industry has several sources of turbulence. This is reflected in the institutional structure of the industry, which in Norway is adapted to handle turbulent change. Examples of this embedding of risk may be found in the organization of fishing in a family and community context and in the regulations of fishing where social objectives are defined at the policy level.

Regulation of fisheries based on resource management policies may represent a source of turbulence. On this basis, crises related to the legitimacy of resource management policies and regulations may result. Because of problems of control, effective resource management policies demand some level of industry co-operation. Development of co-operation in resource management policies may be obtained through different strategies. One strategy focuses on decentralization of management competence through 'co-management'. The hypothesis presented here takes another point of departure: co-operation may be obtained through the definition and implementation of social objectives.

This hypothesis refers to a general relationship between social objectives and turbulence, where a social objective is considered as a basis for a social contract which gives those people affected by resource management policies certain rights – in exchange for their co-operation. A number of Scandinavian studies have explored the relationship between 'dialogue and development' in industry. In this respect, therefore, Scandinavian experience, in which social objectives are instrumental in industrial restructuring processes, are relevant to the discussion. More generally, the question bears upon some familiar issues in the social sciences, namely collective action and order. The solution to such problems may be found in institutional theory.

The problem of order

How is order created? Several answers are possible, but the scope of the present chapter does not allow elaboration of these answers. Let me, therefore, start with a brief statement: order is created through institutions. This does need some elaboration.

Although the concept of institution has become very popular, particularly through the increased interest in the so-called socio-economic perspective, there is no consensus as to its definition. Furthermore, much remains to be said on the question of order and chaos, particularly if one looks at societies which experience swift processes of change.

Giddens (1979; 1984) defines institutions as routines and practices which structure society through the binding of time. Institutions are modes of signifying and allocating resources and authorizing their use which are relatively stable over time. In this way, institutions help to give the world meaning; they establish rules for handling resources and they define the values to which we submit.

Stability is created by a deep human need to preserve our belief in the world as a safe place – our 'ontological security'. Stability is achieved through the implementation of institutional practices; for example, industry might look to the stabilizing influence of institutions when faced with the instability of the market. Examples of this are pointed out by Granovetter in his discussion on transaction cost theory. Transaction costs may be reduced by embedding relations in social institutions (Williamson, 1975; Granovetter, 1985; 1992). A frequently used example is the family, which is often recognized as an effective and stable basis for handling the turbulence and flexibility of small firms.

Another source of institutional stability is formal organizations. As indicated by Williamson (1975), transactions inside formal organizations are regulated by fiat and may therefore proceed more easily than market transactions. A third source of stability is regulation by the state or by other international organizations. Regulations, enforced by the authority which defines them, establish stable elements which are necessary to actors whose destiny depends on the market.

Through the balance between stability and instability, we may sustain turbulence and change without losing our perception of the world as safe and, at the same time, cope with elements in our environment which are extremely uncertain. Giddens has pointed out that we live in a 'high risk society'; our economy depends on sources of funding which are uncertain; we travel by cars and planes which may suddenly be violently destroyed – and so on. This high risk society has developed a specific kind of logic, based on the calculation of risk. If the risk is perceived as small enough, we ignore it. This calculation of risk depends on another factor – a reference system of stable institutions, in which we trust (Giddens, 1991). A number of authors have pointed out that there is a balance between risk and trust in these relations: the higher the risk, the greater our need for trust in institutional structures.

In times of change, this balance of forces between stability and turbulence may be altered. Economic restructuring, institutional reforms or other kinds of deep change may threaten or destroy the institutions which stabilize our lives. What happens then? Giddens' answer to this question is that an actor submitted to deep change will have to go though a period of reflection in which she or he will try to reconstruct a new reference system with a new balance between stability and turbulence. This attempt may have two different results.

A successful reconstruction of a new reference system depends on the remaining elements of the old system. Reconstruction in that case means that the actor places higher trust and relies more heavily on the remaining elements. If you lose your job, you may rely on funding from your family or from the social security system. If you have elements which survive the transition, you are able to act rationally and develop a new reference system adapted to the new situation.

In times of deep change, a reconstruction of the reference system might by interrupted, because all the relevant elements are gone. In that case, according to Giddens, the players in the system may find their ability to act rationally is disturbed. They may be paralysed or may act by trying to establish new groups which give stability – through aggressive behaviour. This statement may be debated. However, if we look around us we will find a number of examples which seem to support this theory. Deep institutional change may lead to processes where people are actually deprived of their capacity to act in a constructive and 'rational' way. They are left with strategies where stability is restored through the formation of groups based on aggression towards outsiders and destructive action.

Assuming Giddens to be right, what are the implications of this for organizing transition and change? The implementation of industrial change should be balanced by policies which preserve other elements of the reference systems of the actors which are impacted by the change. Change in one area of life must be balanced with stability in other areas – and the stable elements may form the basis for the actors' adaptations to the new conditions in their lives. Policies for resource management may destabilize the lives of a number of fishermen, industrial workers and others in fishing dependent regions. This destabilization should be balanced by social objectives, which stabilize some of the central elements of the reference systems of these people.

Social objectives may refer to other elements of an individual's or community's reference systems apart from those directly influenced by resource management policies. In that case, an argument may be made that social objectives facilitate the implementation of resource management policies and reduce their social costs. This hypothesis leads in the direction of arguing that, for fisheries management, social objectives must be defined through analysis of the reference systems of the people living in the coastal regions. However, as North (1990) so powerfully argues, the institutional structures vary considerably between different regions.

It may seem that the Norwegian experiences during the last decade, with regulations which were imposed based on resource management policies, are relevant. Symes & Crean (1994) observe that institutional structures of the coastal regions in the EU differ from those in Norway. European fisheries are incorporated within what Symes & Crean call 'capitalistic systems of production', thus creating an impression of urbanized fisheries, organized in a manner similar to an urbanized agro-industrial complex. In Norway, on the other hand, the fishing industry is mostly located in rural villages and small towns and predominantly embedded in rural institutions such as family firms and rural communities. Thus, the institutions embedding the economy of rural Norwegian coastal regions may be quite different from those of the urban European fisheries.

Scandinavian studies can provide insights into urban industrial restructuring based on traditions of co-operation, in which social objectives are defined and social policies implemented as an integral part of the industrial restructuring process. Two empirical looks at industrial restructuring are presented in the following pages. The first looks at industrial rationalization in the steel town of Mo-i-Rana and the second at the implementation of resource management strategies in the Norwegian fishing industry. In both cases the definition of social objectives is integral to the process of restructuring.

Industrial restructuring

Mo is the administrative centre in the municipality of Rana, which corresponds to the daily commuting area around Mo. The town is situated just south of the Arctic Circle in the northern region of Norway. Iron ore was first extracted in 1902 by the English company Dunderland Iron Ore Company Ltd. The Norwegian Government took control of the mines, nationalized the steel industry and built an integrated iron, coke, steel and cylinder works during the period 1946–65, known as the Norsk Jernverk AS. Population increased from 9000 inhabitants in 1946 to 26 000 in 1972. The work force came mainly from the surrounding rural district where fishermen and farmers dependent to a large extent on a subsistence economy were recruited as industrial workers. In 1981, 3910 of the area's inhabitants were directly employed by Norsk Jernverk AS. In 1980, the industry accounted for 49% of the jobs in the area. The nationalized industries managed considerable technological skill through the employment of qualified engineers and skilled labour.

Norsk Jernverk AS in Mo was an integrated, state-owned consortium which, until 1988, was directly controlled by the Government through the board of directors chaired by the Minister for Trade and Industry. Both production and consortium leadership became deeply rooted in the local community. Executives and employees had strong ties in the town through work and family. The executives of the consortium were, therefore, loyal both to the local community and the Ministry of Trade and Industry as owners. The consortium's organization was characterized by a strong bureaucracy and central leadership combined with employee participation. Employees also had direct representation on the board in accordance with state law.

Until 1988, the ironworks in Mo had the largest branch within the Norwegian Iron and Metalworkers' Union. The trade union movement had 8000 members in a town with 25 000 inhabitants. In the post-war period, both the federation and the local union have influenced the development of national wage levels, the forming of agreements and the introduction of new reforms in the working conditions. Shop stewards at the ironworks were recruited as foremen, administrative employees and staff managers. The Labour Party, local council and federation recruited members from amongst shop stewards in this government consortium. Mo-i-Rana and Norsk Jernverk AS had therefore a close network of social democrats in positions as owners, managers, trade union leaders and mayor.

In 1987 the Ministry of Local Government and Labour instructed the consortium to set up a new plan for restructuring. The proposal was strongly opposed by both the trade union and local industrial leaders. The union's aim was to maintain an integrated state-owned production system in the municipality, securing jobs for everyone. The government plan for privatization would lead to the loss of 2000 jobs and an extremely uncertain future. Only the consortium management and employers' confederation were positive towards the plan.

Trade union leaders in Rana changed their tactics in their dealings with the Labour Government. Opposition to the proposition was used to bind the Government to obligations concerning the local community when the restructuring process began. In the negotiation of these obligations the union held a very strong position. 'Restructuring' was gradually redefined from a disaster to a short-term effort which demanded sacrifices, but which also opened up the opportunity for rewards.

The negotiations resulted in a compromise between the Government and the union approved by the *Storting* in June 1988. This compromise consisted of several elements including: payment of NOK 500 million in a reorganization fund to the Rana municipality and NOK 1 billion for restructuring the consortium; special employment schemes for those made redundant and early retirement opportunities for those over 60; and the promotion of new industrial activity – all in return for the winding up of unprofitable segments of the steel industry at a cost of 2000 jobs.

Union and industrial leaders regarded this compromise as an optimal solution. The union had used its strong political relationship with the Government to force an acceptable compromise.

In 1988, the process was interpreted as a *dugnad*, or time limited common task, demanding sacrifices from everyone and promising rewards through a compromise. The values of equality and individual sacrifice for the benefit of all were strong. The local identity, defining oneself as a member of the union and as an inhabitant of Mo, was well developed in the workforce, as well as among other citizens of Mo. When restructuring was put on the agenda, the experience and value of solidarity co-operation through the concept of *dugnad* was well established. The financial resources put on the negotiating table by the Government made the process tempting for the union. By including the local trade unions and local authorities in the work to develop new jobs and allocate new benefits, support, loyalty and effort were encouraged.

The changes were organized and carried out in a way that meant that new phenomena could be interpreted and managed within the framework of familiar paradigms and by actors with local legitimacy. It was possible to interpret the re-allocation of rights to jobs within the framework of the traditional values of those affected. Important in this context was the principle of seniority – a central issue in traditions. Older employees were given the first option of new jobs at the expense of the younger employees. This decision was generally accepted and helped to legitimize the process of reorganization. The fact that traditional values relating to the right to work and to practices of working life were being observed helped to maintain the image of equal exposure to external means. Faced with the legitimate and objective criteria, which regulated priority in the allocation of jobs and welfare benefits, everyone was equal. The process could thereby lay claim to support and co-operation from everyone.

The distribution of benefits was managed and supervised by known and legitimate local parties; namely the unions and local industrial leaders. In that way, the parties tended to interpret what was happening as being in accordance with the value of equality. All were equals, given the right to new jobs and welfare benefits which were distributed according to principles and procedures which were accepted through negotiations.

Workers who lost their jobs left without protest. The world was changing, but the change was comprehensible and interpretable within principles which corresponded to the widely respected traditional values and laws of the established industrial system.

Fishing

The fishing industry experiences several sources of turbulence. Fishing may be regarded as 'industrialized hunting' with high risks; the results depend not only on markets, labour relations, competition and the other sources of turbulence, which are found in any mainland industry, but also on other sources of uncertainty related to the finding and catching of a prey hidden deep in the waters of the oceans, and based on fluctuating ecological systems. This extreme turbulence is reflected in the institutional structure of the industry which, using Norway as an example, is well adapted to handle change. Examples of the embedding of risk may be found in the organization of fishing in a family and community context.

Coastal 'peasant' fishermen have adapted to the production flexibility of a seasonal economy and the functional flexibility of a coastal labour market through households in which women work in the service industries or the fishing industry according to demand. Through such flexible households, with a capacity to combine different sources of income throughout the year, the economy has adapted to the seasonal flexibility of nature. This seasonal flexibility is also maintained through low capital investment and reducing capital costs. The main effects of the crises following the regulation of the cod fisheries was a destruction of that part of the economy with the highest capital costs. At the same time, ownership of shore-based industry became concentrated.

A major concern in these regions is linked to the frequent exits of young people who have come to regard their parents' work as too uncertain. Artisanal fishermen depend on political support to protect their rights of access to resources, where capital intensive fishing is an aggressive competitor. The peasant economy is in conflict with the pro-moters of industrial modernization. In fisheries, family firms are important structurally. The turbulence of fishing is balanced by the flexibility of the family. Artisanal family business communities have one important characteristic: they depend on the biological reproduction of their populations and on their social and cultural reproduction through socialization of the families into community life. The moral economy of the communities and the technology of their business operations are open only to insiders born in the communities. This creates problems. For stagnating peasant communities, the low recruitment of young people into the positions of their parents is a problem, leading eventually to demographic stagnation. For the successful family business community the lack of labour is a dominant problem.

A number of authors have noted that labour intensive fishing, embedded in a household economy with other sources of income such as a wife working in a public sector job or running a farm, is a system with a large degree of structural flexibility capable of handling most ecological fluctuations. With a modern and well equipped fishing boat costing around NOK 1–1.5 million, it is still possible to invest in a fishing boat and pay for most of the investment during a decade of good fishing. However, the 1980s did not prove to be such a good decade.

Peasant systems are clustered in specialized communities, where fishermen enjoy the advantages of agglomeration economies through reciprocal community relations with their colleagues. These communities are constituted through reference to the fishing technology which is the basis of community specialization. However, reciprocal rela-tions are not necessarily limited to one location. There are reciprocal relations between communities in Lofoten, communities of fishermen further south who come to Lofoten during the winter season, and communities further north in Finnmark which receive fishermen from Lofoten during the Finnmark season. Through these relations, fishing is organized and new technology is distributed. Communities also embed the recruitment of crews and the complex organization of different types of supplementary support activities, such as preparing and repairing equipment. The operative relations during fishing are reciprocal rather than market related. These communities have their own pools of 'core technologies', based on experimental skills, which are only available to insiders.

Within such communities we find a functional flexibility within what Seierstad (1985) calls 'the coastal labour market'. In this labour market, men adapt to seasonal varia-tions through fishing, construction, transportation and agriculture. Increasingly, public sector jobs and jobs related to tourism are being included in these flexible systems. In capital extensive coastal fishing, fishing is an integrated part of households along with a number of other income opportunities, such as public sector jobs, tourism or welfare benefits.

The crisis during the second half of the 1980s led to a contraction and a consequent 'increased peasantization' of existing fishing. Crews were laid off, because the owner could not afford the crew. To the vessel owners, laying off crew was a moral problem because their relations to crew members were reciprocal on a community basis. Boats were partly household operations as a result of the recruitment of the owner's wife as crew, or through co-operation between owners on a community basis where owners manned each other's boats sequentially during the few days each boat was allowed to fish. Through contraction and 'peasantization', where fishing was 'subsidized' by other activities through the household economies of fishermen and fishing boat owners, the main fishing boat owners were able to stay in business. However, crew members, young people or others in more marginal positions were forced to quit (Mariussen *et al.*, 1990). The crisis was paid for through increased public funds for the support of the unemployed.

Regulation did lead to political conflict. One line of conflict reflected regional divisions among fishermen, concerning who should cut back on fishing. However, the main criticism was not directed against regulation as such, but tended to focus on the fact that social objectives were not being met as a result; particularly those social objectives linked to maintaining the coastal population and the small fishing communities. In that way, the agenda for debate focused not on regulation itself, but rather on what was perceived as a crisis for the rural fishing communities (Sagdahl, 1992).

Changing the agenda meant that pressure was removed from resource management policies, as discussion concentrated on measures to preserve and develop the communities. The development problems of fisheries dependent regions were analysed and discussed by central municipal authorities (Mariussen *et al.*, 1990; Onsager, 1991; Onsager & Eikeland, 1992, Mariussen *et al.*, 1994). Both central and regional authorities were engaged in plans for development of alternative industries, such as tourism, to be implemented by regional authorities and based on central government funding. At the same time, social policies helped to secure rights for support to the unemployed while labour market authorities implemented re-training programmes and offered education opportunities for the unemployed.

As a result of these policy measures, together with the flexibility of the household economy, the regulation crisis in fisheries did not seriously impact on the general social welfare of the population in the fisheries-dependent regions. The standard of living and level of social welfare in regions affected by the crisis is today not lower than the national standards (Fylling *et al.*, 1994). Thus one may argue, social policy, labour market policy and regional development policy measures helped to facilitate the implementation of resource management policies.

Social objectives as social contracts

In the two cases discussed above, social objectives are of different types, namely:

- institutionalized regulations and welfare systems, giving rights according to universal criteria which apply for all;

- policy objectives, such as the population and settlement objectives of the Norwegian fisheries policies, defining maintenance of rural fishing villages as a policy objective;
- *ad hoc* negotiations with the organizations involved.

In the case of Mo, the position of the union was strong. The initiation of negotiations on restructuring was based on a long tradition of negotiation and co-operation in questions of industrial relations. The privatization process was implemented by the leaders of the established institutions of the town, who used the moral and political power of these institutions to transform them. Privatization was collectively reinterpreted as a common effort to restructure and in that way save the industry of the town. In the process of re-interpretation the rituals of the established system of industrial negotiations were used and loyalty to the town was mobilized.

To the people involved in the process, this embedding of change in established institutions implied an element of continuity and stability, balancing turbulence and uncertainty of change. The element of stability depended on the social objectives which were formulated as central elements of the compromise.

The structures of fisheries-dependent regions vary. In the case of resource management regulations, the population patterns objective was not put on the agenda as a question for formalized negotiating of resource management policies. However, in the public discourse over these policies, the social objective was referred to. The debate on resource management policies therefore turned into a debate on threatened fishing communities. This change of agenda opened up demands for policies to preserve fishing communities through a number of social policy measures outwith the area of the fishing industry itself. In that way, criticism of the resource management policy soon faded, as emphasis was put on the communities' capacity to diversify and develop other industries. Here, the growth of tourism in a number of fishing villages was an important vehicle. The stability of the villages – and the social network of the welfare state – helped those involved to cope with the instability of fishing.

A proper definition of a social objective must take into consideration the reference system of the people involved. An efficient social objective should facilitate implementation of resource management policies, by compensating the increased uncertainty that the policy imposes on the people living in coastal regions with an element of social security. This element may or may not be related to the fishing industry. Social objectives may be defined on several levels:

- *The region or locality.* A social objective may refer to a region or a locality, as for instance the Norwegian population pattern objective.
- *The household.* A social objective may refer to the household of the people involved. Both in Norwegian coastal societies and in European industrial regions going through processes of restructuring, the household becomes an important element in coping.
- *The individual.* A social objective may be defined as individual rights to certain welfare services, as referred to in the example of Mo-i-Rana above, where welfare payments were granted, depending on certain criteria.

Social objectives must correspond to values of the participants in the process and the institutions in their regions. These very institutions must be expected to vary considerably between different regions. The identification and measurement of social parameters for the definition of social objectives will, therefore, prove to be a complex task. The kind of 'social data' needed for a successful development of social parameters would seem to imply both universally defined indicators, as well as culturally and regionally specific variables. To define these parameters properly, institutional analysis of the various coastal regions is imperative.

It is implicit in this contribution that comparative research into processes of industrial restructuring, where social objectives are incorporated in the planning and implementation of the restructuring process, is highly relevant. It is, moreover, clear that successful use of social objectives as a basis for social contracts, facilitating the implementation of resource management policies or any other industrial restructuring programme, is not uncomplicated. Indeed defining social objectives which fit the reference systems of the fishermen and industrial workers affected by new resource management policies is a profoundly complex task.

Social objectives, properly defined, can increase the element of local co-operation in the implementation of resource management policies, facilitating the implementation of new regulations and increasing their efficiency. Effective implementation of properly defined social objectives may prove a cost-effective way of reducing the social costs of resource management policies.

References

Fylling, I., Hanssen, J.I., Sandvin, J. & Størkersen, J.R. (1994) *Vi sto han av!* Report from Nordland Research Institute 16/94, Bodø.

Giddens, A. (1979) *Central Problems in Social Theory.* Macmillan Press, London.

Giddens, A. (1984) *The Constitution of Society.* Polity Press, Cambridge.

Giddens, A. (1990) *The Consequences of Modernity.* Polity Press, Cambridge.

Giddens, A. (1991) *Identity and Self in High Modernity.* Polity Press, Cambridge.

Granovetter, M. (1985) Economic action and social structure: the problem of embeddedness. *American Journal of Sociology,* **91** (3), 481–510.

Granovetter, M. (1992) Economic institutions as social constructions: a framework for analysis. *Acta Sociologica,* **35**, 3–11.

Mariussen, A., Muller, H. & Høydahl, E. (1994) *En Fortelling om to Kystregioner.* Report from Nordland Research Institute 13/94, Bodø.

Mariussen, A., Onsager, K. & Tønnesen, S. (1990) *Fiskerikrise – Virkninger og Omstilling i Kystsamfunn.* NIBR notat 134, NIBR, Oslo.

North, D.C. (1990) *Institutions, Institutional Change and Economic Performance.* Cambridge University Press, Cambridge.

Onsager, K. (1991) *På Rett Kjøl i Åpent Farvann?* Regionale trender nr 2.

Onsager, K. & Eikeland, S. (1992) *Ny Giv i Nordnorsk Kystindustri.* NIBR rapport 8, NIBR, Oslo.

Sagdahl, B. (1992) *Co-management, a Common Denominator for a Variety of Organizational Forms.* Report from Nordland Research Institute, Bodø.

Seierstad, S. (1985) Arbeid som rettighetsgivende status i et kapitalistisk arbeidsliv. *Tidsskrift for Arbeiderbevegelsens Historie 1.*

Symes, D. & Crean, K. (1994) Social issues and the socio-economic paradigm in fisheries management. Theme paper for workshop entitled 'An Agenda for Social Science Research in Fisheries Managment'. The Borschette Centre, Brussels.

Williamson, O.E. (1975) *Markets and Hierarchies.* Free Press, New York.

Chapter 4
Community Fish or Fishing Communities?

AUDUN SANDBERG

Introduction

Mounting pressures on the Common Fisheries Policy (CFP) of the European Union take many forms. Together these pressures should make whatever uniform fisheries policy the Union might have had, crumble long before the year 2002:

- empty quotas ('paper fish') for the North Sea;
- serious deterioration of European coastal habitats;
- destabilization of the European fish market, following the collapse of the Russian market at a time when local landings were low and prices should have been high;
- over-investment in distant-water fishing fleets at a time when distant waters have been increasingly closed off by coastal states all around the globe;
- threat of organized crime filling an institutional vacuum following the eventual introduction of the CFP in the Mediterranean region;
- socially motivated over-investment of structural funds in fisheries-dependent regions.

With such bleak prospects, it is worth noting that social scientists have a tendency to concentrate their analysis on what has brought about the destruction of traditional systems of managing a resource. This might yield some valuable insights into the mechanisms of past social change, like the transition from *gemeinschaftlich* to *gesellschaftlich* society. But the real challenge for modern social science is to study what it is that does not work in the purpose designed resource management systems in the age of 'high modernity' (Giddens, 1991). It is necessary to take up this challenge and to ask what social scientists can offer in terms of constructive analysis for a new or a revised CFP.

All the pressures mentioned above are part of the 'crisis' of European fisheries. According to Schumpeter (1934), it is during such times of crisis that inventive restructuring is accomplished. Now is the time to change the basic structure of incentives through conscious collective action: but now is also the time when entrepreneurs can grasp the opportunity created by the misfortune of others – non-competitive actors – and shape the future for themselves. To the extent that the CFP produces an institutional vacuum at the local level, it is within this vacuum that entrepreneurs can shape the future, so that the year 2002 becomes a *fait accompli* and there is no going back to traditional ways, nor to the image of a purpose designed Common Fisheries Policy. In such a fluid situation the social scientist tends to objectivize the fisherman; his strategies

34

are often analyzed and interpreted both by economists and other social scientists within the current political environment and the current incentive structures. Who is 'efficient' and 'competitive' is always relative to the rules of the game, that is to the institutional structures of a given society and a historical epoch. Therefore, an analysis of the formation of a new fisheries policy also requires objectivization of the social scientists; what are their interests, their frame of reference, and their ability to see beyond the prevailing institutional establishment (Bourdieu, 1992). More often than not, fishers can, individually or collectively, offer self-reflections that go beyond the theoretical paradigm of the social scientist but, being trapped in the struggle for daily income, they cannot act otherwise. The professional and social duty of the social scientist is to explain the relations between the institutional set-up and the outcomes of the actions of fishers.

One such fundamental relationship is identified by asking the simple question: for whom are the fish resources common? If they are common to the European Union as a whole, the resulting institutions will produce a certain set of strategic choices on the part of fishers. If it is common to a nation state or to a region within a state, the resulting institutions produce different strategies from fishers. If the resource is common only to one or several coastal communities, this again implies very different social institutions and very different strategies and outcomes of the actions of fishers. By asking this simple question, one can avoid the fallacy of a seemingly necessary connection between a Common Fisheries Policy and a Common Pond for the EU. Logically there should be no contradiction between a Common Fisheries Policy and institutional arrangements that allow fish resources to be common to smaller units within the Union, for example regions and coastal communities. A Common Fisheries Policy can also be built from below, from a multitude of coastal cultures and institutional arrangements.

The historical development of fisheries management

Fishing activity has always been dependent on politics. During Roman and medieval times, the channels of influence from fishers to emperors and kings to clerical and feudal lords shaped the governing conditions for establishing fishing harbours, for using forests for boat building and for obtaining privileges necessary for the marketing of fish. It can be argued that politics has shaped the entire pattern of fishing hamlets along the coasts of the Atlantic Ocean, the Mediterranean Sea and the Baltic Sea, out of a concern for the relations between fishers and other agents in society. For instance the whole northern part of the sea between Norway, Greenland, Iceland and Scotland remained a prohibited area with *farbánn* for foreign fishers and traders from the Viking age until the liberalization of the eighteenth century. Initially this was because it was in the king's interests; later because it was in the interest of the Hanseatic League. What we today call traditional fisheries have been shaped by the political history of Europe.

After the liberalization of fisheries in the eighteenth century, following the Dutch doctrine of the 'freedom of the seas', the advancement of sophisticated fish finding and fish gathering equipment brought about a new relationship that became a major concern of the state: the relationship between their national fishers and a limited and productive resource which could be exhausted. While through the centuries marine resources were believed to fluctuate and migrate haphazardly, the state now saw a new role for itself, that of managing the resource directly, with a species-specific maximum sustainable yield to be managed for the benefit of the balance of trade and needs of the national treasury. But the stock-properties of real life were different from the neat equilibrium models; and both states and intergovernmental organizations experienced fundamental problems in securing a steady flow of each selected species of fish from the managed stocks of European waters. Especially on the highly productive continental shelves of the higher latitudes, where the ecosystem tends to be basically unstable, it is natural for dramatic changes to occur; indeed this is one of the main reasons why these ecosystems can yield such enormous amounts of fish. But perhaps rational and sophisticated multispecies modelling and management techniques could secure at least a steady flow of some sort of commercially valuable fish. The concept of maximizing the sustainable yield was thus switched from a number of preferred species to any combination of species. However, except for a fragile agreement among biologists and fisheries managers that a large and stable stock of herring is the backbone of any viable multispecies ecology, there is dwindling scope for state management of the seas for a stable maximum yield and maximum economic and social benefits for coastal communities. In many respects, it is the continued official belief in a management rationale that has not – and probably never can – produce a predictable future, that is the deep cause of the social crisis in many European fishery-dependent regions.

When the present resource management system is depicted as a confluence between this obsolete rationale of a manageable future and the continued preservation of the politically convenient relative stability of quotas, it becomes possible to identify the strong social forces that work towards an institutional breakdown in a multitude of coastal communities. The logical strength of these forces is of such a magnitude that it warrants the question: is there an inescapable choice to be made between 'community fish' or 'fishing communities'? If one chooses the one alternative – a 'common pond' for the whole European Union – one cannot simultaneously choose to have 'fishing communities'. And if community fish is the choice, are there feasible designs of management regimes that can work without fishing communities? If it is technically possible to replace fishing communities with company or industry designs, do these have lower transaction costs than the fishing community design? And finally, is this line of development socially desirable?

Property rights: a system of privileges

To shed some light on these fundamental questions, let us explore the question of property rights, which is considered by many to be the basis for all incentive systems (North, 1991). Here, there is massive conceptual confusion – also among academics – which tends to blur the debate on necessary institutional changes. The theory of common property requires us to specify to whom the property is common, in other words who belongs to the specific group of proprietors with certain rights and duties towards the resource (Ostrom, 1991). If the resource is common to everyone – a 'Community or Union pond' – it is really a public property where no group of proprietors have any rights and duties towards the resource; and public property is not a commons. It is then the nation state or the European Union which is the real owner and which can issue access rights and harvest rights to state-authorized individual fishers and authorized fishing companies. The property rights are thus not privatized at all; they remain in the public realm – the realm of the state. It would therefore be more correct to define a system of historically stable quotas extending all the way down to the individual fisher level as a system of privileges – the privilege of state protection for some fishers against other potential fishers. In some countries the distribution of such privileges is relatively stable, in other countries with a fully developed system of individual transferrable quotas (ITQs), they can be accumulated, transferred or abandoned. Thus the strange combination of community fish and 'historically stable quotas' in many respects implies 'a refeudalization' of the coastal areas of Europe.

Part of the social crisis in European fisheries is the decreasing value of these privileges. Because of the conceptual confusion, quotas have to some extent been treated as quasi-property rights (or 'imitated property rights') and have been entered as securities for loans far above the real financial value of the privilege. In much the same way as privileges became empty under the threat of state bankruptcy during the decline of feudalism (North, 1981), the privileges of fishers have gradually been eroded. This is one of the factors that undermines the purpose-designed resource management system of the age of 'late modernity'. There are two ways of analysing this erosion:

- One is linear analysis; when there are too many privileges issued relative to the size of the resource, each privilege loses its value and the holder risks financial bankruptcy. The linear solution is to reduce the number of privileges, that is reduce overcapacity so that the remaining privilege holders can make a decent living. This means using interventionist instruments to reduce the fishing sector even more and actively remove fishers from harvesting activities. This reduction of the fishing communities runs the risk of drying up the professional fishing culture of the fishing communities so that they get progressively 'thinner' at each successive downturn in the natural stock fluctuations, and finally disappear as active fishing communities.
- The other is a dynamic analysis of the function of the protective element in the privilege itself, and the relationship between this and the basic incentive structures. Despite a massive display of micro-economic engineering effort in the construction of quota systems, the state has not been able to protect its fishermen against other fishers on the seas or through the markets. In addition, the quota system has created an institutional vacuum at the local level that renders the fishers unable to protect themselves. The privilege gives no protection against other fishers who will always work through the market. Thus the value of the protection element in the privilege is also eroded, and only a steadily increasing amount of control effort and substantial financial support – with mounting public expenditure for the benefit of a dwindling number of fishers – guarantees the temporary survival of the system of state property rights and fishing privileges.

Taken together, the absolute uncertainty of the financial value of the privilege and the erosion of the protective element of the privilege undermine both the internal and the external legitimacy of the 'refeudalized' system. It is surprising also that in periods of upturns in certain fish stocks, like in Arctic cod in recent years, the substantial earnings made by the reduced number of privilege holders tend to reduce the legitimacy of the system. Unemployed youth in coastal communities cannot accept the closure of fishing and the limitation of 'super-earnings' to a few, while there obviously would be sufficient fish to give all a decent share. Some sophistication of the quota system – like the Norwegian 'recruitment quota' and 'periodic group quota' – can soften the social effects of a rigid privilege system, but these will always remain inferior alternatives to the individual fisher as long as a vessel quota or ITQ is in force.

It is important to understand that it is the permanent character of even imitated property rights that creates the rigidity. Experience shows that once a quota system is in place, it is very difficult to add new quota entitlements for a species of fish which has an upswing in abundance, in order to create more employment in a particular fishery. Existing quota holders will claim that they are justified in keeping a good year's catch for themselves as compensation for all the poor years in the past and, maybe, also in the future. In the same way, it is very difficult, in the short run, to take away existing quotas from fishers who may have invested on the basis of what they thought were secure harvesting rights. The intransigence is increased even further by the various rules imposed by the different states on their privilege holders; very often participation in the poor year's fishery is a prerequisite for the extension of the privilege into future and richer years.

Re-defining fisheries policy

If both the idea of a 'common pond' and the idea of 'historically stable quotas' – and especially the combination of the two – do bear a heavy responsibility for the current social crisis in European fisheries, what should then be the alternatives open to decision makers and designers of European institutions? Would a clearer definition and a recognition of fishery-dependent regions within the Union's regional development programmes provide a more positive role for state or Union intervention in fisheries than the present attempts at unitary regulations for the whole CFP area; should the Union adopt a policy that acknowledges and encourages social and institutional diversity? Or, if the erosion of the Common Fisheries Policy continues, should the coastal communities not await new initiatives from the Union, but move instead towards reclaiming the coastal commons by various political and organizational strategies?

To give a clear answer to such fundamental questions requires penetrating analysis, and it should be pointed out that these are not simple one-dimensional questions which can be satisfied by one-dimensional answers. Some aspects of European fisheries policy, such as pollution control, total allowable catches for pelagic or migrating stocks of fish, the institutional framework for a smooth marketing of fish, etc., have to be determined at an international level. Other aspects of European fisheries policy, like the distribution of fishing rights, the recruitment to fishing and retirement from fishing, can most efficiently be handled at the local level.

One immediate answer would, however, be to point out the need to pause and think twice about pressing ahead with a new round of uniform regulations from 2002, pending a more thorough analysis of the consequences of the incentive structure inherent in the present CFP.

As a modest contribution to such an analysis, we shall look briefly at some aspects of the incentives that constitute the institutional design of European fisheries. At the risk of making sweeping statements, we shall treat the CFP of the European Union, the basic regulatory institutions of the Nordic countries and of the Eastern European countries in the Baltic Sea as containing basically the same kind of incentive structures: national or 'federal' common seas, extensive mobility for fishing vessels and stable historical quotas extending down to the individual fishing unit.

The structure of incentives

One of the basic requirements of a balanced incentive structure is that there is an approximate correspondence between the *rights* of fishers to harvest in the coastal areas of Europe and the *duties* European fishers have towards maintaining the productive capacity of coastal waters and the supporting social infrastructure of the coastal communities.

If there are substantially more rights than there are duties, the fish resources are likely to be exhausted within a short span of time. Distant-water fishing near the shores of other fishing communities exhibits this type of fishing rights without accompanying duties.

If there is poor correspondence in the distribution of duties and rights, the resource management institutions will crumble from within because of a lack of legitimacy. Here fishers will often 'take back their rights', and 'black fishing' and 'black trading in fish' will flourish and the cost of government control will mount. One way of analysing the lack of correspondence of rights and duties is to subdivide what we conveniently call property rights into its five distinguishable elements (Sandberg, 1993):

- the right of access;
- the right to harvest;
- the right to manage;
- the right to exclude; and
- the right to alienate.

The imitated property right of the conventional individual quota is then a bundle of rights that contains only two of these rights; the right of access and the right to harvest, and these are counterbalanced by the 'minor' duties to behave properly on the fishing ground and refrain from overfishing. By contrast, the crucial duties for the survival of coastal communities are mostly contained in the rights to manage and exclude.

These are vested in a national or a federal governing body which has to spend increasing amounts of resources on ensuring that fishers do not exceed their limited rights. From empirical data, we now learn that the institutional arrangements with the best correspondence between rights and duties, are those based on collective rights, where the group to which the resource is common is not universal, but limited and bound to each other in some form of network of obligations or social contract – and one which also has rights to exclude and to manage. Entry and exit is possible, but is regulated, often with a mandatory initial period as a novice – like the *cofradías* of Spain. Compared to a government-run system with state-authorized fishers, such collectives tend to have considerably lower transaction costs and would therefore in the long run provide more efficient institutions (North, 1991).

A second important part of the incentive structure is the temptation and opportunity to protect the fishing profession from newcomers and intruders. This has to be balanced by the incentives to secure new recruits to the group of fishers and to maintain necessary dynamic social processes in the coastal communities.

If the degree of protection achieved by state-authorized fishers becomes too strong, recruitment will suffer; coastal communities will become rigid and vulnerable and fisheries will lose legitimacy as an important employment factor on the coast. If, on the other hand, the degree of protection from intruders is too weak and recruitment becomes too large in a Europe of mass unemployment, the social conditions of fishers will rapidly decline and the 'poor fisher' will again be a common category in coastal communities. And poor fishers tend to fish harder when prices become lower, thus accelerating the fluctuations of the fish stocks even further.

A third part of the incentive structure is the temptation and opportunity for various groups of fishers to be flexible and/or mobile in their fishing operations. Until recently there was a balance between the extremely mobile, but specialized, ocean fishing vessels and the versatility and flexibility of the coastal fisher. Due to changes in both technology and institutional conditions, the modern ocean fishing vessels have now achieved a high degree of flexibility, while the traditional coastal fisher has experienced dramatically increased rigidity. Both pressures towards capitalization in small-scale operations and the increasingly rigid single-species quota systems have eroded the coastal fisher's earlier advantage of flexibility in harvesting operations. He can no longer switch easily from fish stocks in decline to fish stocks on the increase as was the traditional fishing strategy for artisanal fishers. If it is desirable to continue having coastal fishers and fishing communities in Europe, the basic incentive system must therefore be changed so that the coastal fisher again can reap the advantage of his flexibility. A 'free adaptation' to fishing in coastal waters within a system of regionally defined collective rights to a multitude of species would be one way of reclaiming this advantage in harvesting operations – and to reclaim the coastal commons. However, this would require a partitioned fisheries management regime, with an efficient resource protection of coastal fishers from the highly mobile and flexible ocean-going fishing vessels. For such local and regional incentive structures to work properly, this kind of resource protection, and the necessary control measures, would have to be more efficient than is presently the case with exclusion zones or boxes. Provided an efficient resource protection is achieved, an incentive system based on 'free adaptation' and flexible fishing strategies would require a certain 'overcapacity' in fishing communities, thus reducing the social problems resulting from government interventions aimed at a one-dimensional reduction of the overall harvesting capacity. In sum, a deal between the state and the fishing communities could see the transfer of some property rights to the fishing communities (adding management and exclusion rights to access and harvesting rights); in return, the fishing communities and their constituent households take upon themselves to absorb more of the fundamental ecological uncertainty connected to the harvesting of wild fish. As the state in the age of 'late modernity' is accepted to have limited capacity to deal with people's problems, such a transfer of property rights is a precondition for utilizing the self-governing capacity of local communities.

There are many reasons why such ways of designing incentive structures in coastal communities are more feasible now than at the beginning of the industrialization of fisheries. One is the growth of aquaculture in many European coastal communities, and the development of commercially viable farming technologies for an increasing number of species. To an increasing extent, this will enable coastal communities as a whole to achieve yet another form of flexibility by stepping up the farming of wildstock species which are in decline and actively enhancing the coastal commons, through fry protection zones and other sanctuaries, artificial reefs, etc. (Sandberg, 1991).

But even with an efficient resource protection and more efficient institutions that allow a more flexible use of adjacent resources for the coastal communities, these kinds of incentive structures cannot protect the fishing communities from competition from the world's mobile ocean fishing fleet through the markets. Neither national protection, nor European protection, can avoid the glut of fish at certain times when some natural stocks are in an upswing somewhere in the world's oceans. With an atomistic structure of fishing communities or small, uncoordinated producer organizations (POs), the incentive will be to compensate for a falling price in one species of fish with increased catches of that species. The various compensatory measures, including minimum prices, withdrawal prices and government subsidies for freeze-storage of surplus fish, do not alter the basic incentives; when fishers learn to speculate against this system, it might even amplify the inherent tendency to fish harder as prices decline.

The alternative incentive system, which contravenes the official EU doctrine that POs shall not have a dominant place in the market for fish, would be to allow cartel formation among co-operating POs. At the regional level, POs that are able to pool the quotas of their individual members can today operate more efficiently in the market for fish. Co-operative efforts among a number of POs which would otherwise compete with each other, can channel the correct incentives from the market to the fisher, so that he switches to another species when the price becomes too low. Both for the resources in the sea and the resources in the national treasuries, it would be an advantage that the non-marketed fish is still alive and swimming, rather than withdrawn and destroyed.

Conclusion

The answer to the opening question – is there an inescapable choice between community fish or fishing communities? – is therefore that the continuation of a CFP based on a common pond eventually will produce an entirely industrialized fishery and the extinction of European fishing communities as we know them. But simultaneously it will imply an extinction of the social problems directly related to fishing. If we want to have fishing communities in the future and utilize the capacity that ordinary people have to govern the resources they depend on themselves, there is a number of basic elements in the incentive structures that need to be changed. Such changes will surely meet with intense opposition from organized interests within the industrialized sector of marine harvesting.

But as we have tried to show, it is also possible to craft a Common Fisheries Policy that acknowledges and encourages institutional diversity suited to the multitude of ecological and cultural settings on the European coasts. Such a diversity will offer a management of marine resources that is more transaction cost efficient and has a higher legitimacy than a rigid system of uniform regulations from the Aegean to the Baltic Sea.

It should then be possible to design a more positive role for intervention by the European Union in fisheries. Intervention should aim at a relocalization of management decisions and decisions concerning the design of institutions to the level of the coastal community or fishery-dependent region. This would be a Common Fisheries Policy where the decisions are taken and designs worked out at the level closest to those affected by the outcomes and workings of the institutional arrangements – in line with the original meaning of the subsidiarity principle.

A remaining question is whether such a relocalization requires prior re-nationalization of the CFP. That is a wholly new research agenda and there is no room to embark on it here. But one relatively safe hypothesis is that nation states, which are as vulnerable to pressures from organized industrial interests in fisheries as the Union, are no guarantee for a smooth transition to more localized or regionalized fisheries management.

References

Bourdieu, P. (1992) *An Invitation to Reflexive Sociology.* Polity Press, Cambridge.

Giddens, A. (1991) *Modernity and Self-Identity, Self and Society in the Late Modern Age.* Stanford University Press, California.

North, D.C. (1981) *Structure and Change in Economic History.* Norton & Co., New York.

North, D.C. (1991) *Institutions, Institutional Change and Economic Performance.* Cambridge University Press, New York.

Ostrom, E. (1991) *Governing the Commons: the Evolution of Institutions for Collective Action.* Cambridge University Press, New York.

Sandberg, A. (1991) *Fish for All – CPR-Problems in North Atlantic Environments.* NF-Notat no. 1104/91, Bodø.

Sandberg, A. (1993) *The Analytical Importance of Property Rights to Northern Resources.* LOS i NORD-NORGE Notat no. 18, Tromsø.

Schumpeter, J. (1934) *The Theory of Economic Development.* Harvard University Press, Harvard, Massachusetts.

Part II
Alternative Property Rights Systems

Chapter 5
Property Rights and Practical Knowledge: The Icelandic Quota System

GÍSLI PÁLSSON and AGNAR HELGASON

A powerful 'modernist' paradigm in bio-economics and resource management assumes that the current crisis in many fisheries is due to their open-access nature; thus the popular label 'the tragedy of the commons'. The modernist paradigm also assumes that marine ecosystems are characterized by linear relationships and that establishing the relevant ecological 'facts' is a relatively straightforward task, albeit mainly for 'experts'. The paradigm further suggests that only a 'market' approach, emphasizing commoditization of resources (usually privileging capital rather than labour), will ensure stewardship and responsible resource use. Finally, the issues of ethics and equity are typically rejected as irrelevant externalities or theoretical distractions. Increasing empirical evidence and a growing body of theoretical scholarship suggest, however, that there are good grounds for questioning those assumptions.

This chapter argues for the importance of social research on fishing and fisheries management, focusing on three related issues:

- the social construction of scientific knowledge;
- the nature of practical skills and their role in resource management; and
- the social consequences of individual transferrable quotas (ITQs).

We shall illustrate our general points with reference to the Icelandic demersal fishery. In Iceland an ITQ system was introduced in the cod fishery in 1983 to prevent the 'collapse' of the major stocks and make fishing more economic. This system divided access to the resource among those who happened to be boat owners when the system was introduced, largely on the basis of their fishing record during the three years preceding the system. While originally the system was presented as a short-term experiment, with the fisheries laws passed by the Icelandic Parliament in 1990 it was reinforced and extended into the distant future. Recently, public discontent with the concentration of ITQs and the ensuing social repercussions has been increasingly articulated in terms of feudal metaphors, including those of 'tenancy' and the 'lords of the sea'.

Discursive force and scientific practice

Many economists tend to argue that only the 'hidden' forces of the market can ensure efficiency and sustainable use of resources; for them, privatization is the sole alternative to environmental problems. Such arguments are seductive and powerful in the modern world and there is no need to reproduce them here; for some examples in the literature on fisheries, see Neher *et al.*, (1989). As a consequence, in several fisheries in different parts of the world, fish stocks are being turned into private property. First, the resource is appropriated by regional or national authorities and later on the total allowable catch for a season is divided among producers, often the owners of boats. At a still later stage, such temporary privileges are turned into a marketable commodity.

Many scholars have raised serious doubts and criticisms with respect to the bio-economic theory of privatization and the tragedy of the commons. To begin with, the theory has been challenged on empirical grounds; some privatized regimes are obvious failures and, conversely, some 'commons' regimes function rather well (McCay & Acheson, 1987; Durrenberger & Pálsson, 1987b; Wilson *et al.*, 1994). There are important conceptual problems as well, one of which relates to the tendency to radically separate systems and activities, the social and the individual (Pálsson, 1991). Indeed, the scholarly discussion of property regimes partly hinges on what exactly is meant by terms such as 'commons' and 'individual'. Among students of European history, there is a tendency to blindly adopt fairly recent definitions, viewing such concepts 'through a nineteenth-century – that is, bourgeois – lens, defining them as essences rather than relations' (Roseberry, 1991). Williams (1976) points out that the term 'individual' originally meant 'indivisible', that which cannot be divided, like the unity of the Trinity; the change in meaning, he suggests, the adoption of the modern meaning emphasizing distinction from others, 'is a record in language of an extraordinary social and political history' (Williams, 1976). The concept of the 'commons' has been through a similar treatment. While in medieval Europe it referred to 'community property subject to community control' (Hanna, 1990), nowadays it is frequently associated with 'tragedies' and 'open access'.

More generally, it seems difficult to separate bio-economic theorizing from politics, culture, and rhetoric (see, for instance, McCloskey, 1985; Gudeman, 1986, 1992; Ferber & Nelson, 1993). While scientific knowledge is often conceived as an 'objective' representation of the external world, in reality the scientific enterprise cannot be fully separated from its social environment. McEvoy (1988) contends, for instance, that the thesis of the tragedy of the commons represents a 'mythology' of resource use, a model 'in narrative form for the genesis and essence of environmental problems'.

One example of the rhetorical content of theorizing on enclosure and privatization is the persistent inclination of advocates of systems of individual transferable quotas to privilege capital over labour. Icelandic fishing is a case in point. When the Icelandic ITQ system was first implemented in 1984, each fishing vessel over 10 tons was allotted a fixed proportion (*aflahlutdeild*) of future total allowable catches of cod and five other demersal fish species. Catch quotas (*aflamark*) for each species, measured in tons, were allotted annually on the basis of this permanent quota share. And the fortunate quota holders were the owners of vessels, not crews. This arrangement did not go uncontested, for there have been heated debates about what to allocate and to whom. The issues involved illustrate the discursive contest between different groups of 'producers'. Boat owners argued for 'catch quotas', to be allocated to their boats. Some fishermen, on the other hand, advocated an 'effort quota', to be allocated to skippers or crews. In fishing, they argued, value was created through the application of expertise and labour power and not that of the equipment, the boat and the fishing gear. Catch quotas would be grossly unfair since the 'best' skippers would be assigned the same quota as 'bad' ones. When allocated the same amount of effort, on the other hand, measured in number of allowable fishing days in a season, the 'good' and the 'bad' skipper would catch different amounts of fish. Under such a system of effort quotas, then, successful skippers would be rewarded for their exceptional contribution to the economy by an extra catch.

Fishermen sometimes further insist that as the 'real' producers of wealth they are entitled to quotas. As one skipper put it:

> 'who has more rights concerning quota payments . . ., the man who hires crew men, the one who finds the fish and brings the catch ashore, or the boy who inherits the boat of his father but has never been at sea?'

The allocation of quotas to skippers on the basis of their skills, some skippers have argued, would be economical in the long run; costs and effort might be significantly reduced by making fishing the privilege of the most efficient skippers.

At one point, the authorities partly conceded such criticism when revising the regulatory framework of the ITQ system; boat owners were offered the right to choose between effort and catch. Skippers and crews, however, failed to convince the planners and the authorities that they, too, were legitimate quota-holders. It is difficult to explain the restriction of quotas to boat ownership without reference to discursive force – the cultural assumption that real wealth is 'produced' more by those who are owners of boats and equipment than by the labour power of those who work at sea. Fishermen lost the discursive contest with boat owners and the advocates of bio-economic theory. This contest, and other similar ones, illustrates the need to question the assumptions, concepts, and rhetorical content of privatization and common property theory.

In the Icelandic case, the discursive contest was partly a legal one. During debates on the fisheries laws enacted in 1990, some members of Parliament raised doubts about the 'legality' of the ITQ system, arguing that proposed privileges of access might imply permanent, private ownership which contradicted some of the basic tenets of Icelandic law regarding public access to resources. Lawyers concluded that the kind of ITQ system under discussion in Parliament was in full agreement with the law and that quotas did not represent permanent, private property. The laws which eventually were passed reinforced such a conclusion by stating quite categorically that the aim of the authorities was *not* to establish private, government protected ownership. It seems clear, however, that boat owners have become *de facto* owners of the fishing stocks. The tax authorities have decided, one may note, that ITQs are to be reported as 'property' on tax forms and that the selling of ITQs involves a form of 'income'. Recently, the Supreme Court resolved, in a case between a fishing company and the Minister of Finance, that accumulated ITQs represented private property liable to taxation.

Indeed, the question of who owns the fish in the sea has become a central issue in Icelandic political debate. Boat owners claim that they alone are entitled to the rents produced by the ITQ system. The traditional usufruct rights of boat owners, they argue, should be transferred into permanent 'ownership' of the fishing stocks in the form of a fixed share of the catch. For them, the quota system is only a logical extension of the cod wars and the arguments favoured by the Icelandic Government; a 'rational' use of resources, they claim, can only be expected as long as those who use them are dependent upon them as owners. While in the early stages, ITQ systems only imitate private property rights, later on true property rights, similar to those found in western agriculture, may develop. As Scott (1989) points out, such an evolution of appropriative regimes 'can be expected to continue until the owner has a share in management decisions regarding the catch; and, further still, until he has an owner's share in management of the biomass and its environment.'

In the modern world, ITQs and similar market approaches are increasingly adopted in response to environmental problems. Their wider social and economic implications are hotly debated, however, as they raise central questions of ethics, politics, and social theory. For many of the critics, market approaches to resource management are incompatible with egalitarian sensibilities and communitarian notions of responsibility. Social scientists, including anthropologists and economists, should attempt to examine what the rather loose reference to the 'market' entails (Dilley, 1992). How the narrative of privatization and economic efficiency, and the opposite narrative of the 'public trust' (common use rights), are used in specific ethnographic contexts is an important topic for research.

The nature and role of practical knowledge

Nowadays, with the persistent threat of over-exploitation, fishing is often subject to stringent public regulations and 'scientific' control. Generally, both marine scientists and resource economists present the coastal ecosystem as a predictable, domesticated domain. At the same time there is a tendency to assume that the practical knowledge of those who are in direct contact with the ecosystem on a daily basis is of little or no value for resource management. In Icelandic fisheries management, there is very little attempt to utilize the knowledge that skippers have achieved; skippers frequently complain that marine biologists tend to treat them as 'idiots', reducing practical knowledge and local discourse to mere 'loose talk' (for another similar case, see Vestergaard (1993) on Danish fisheries). There are some signs of change in this respect, one of which is the so-called 'trawling rally' whereby a group of skippers regularly fish along the same trawling paths, identified by the biologists at the Marine Research Institute, in order to supply the latter with detailed ecological information. In some fisheries, including the lobster fishery of Maine in the United States, the relations of power between local fishermen and experts in marine biology seem to be the reverse (see Acheson, 1988).

The denigration of practical knowledge is reinforced by orthodox theories of learning and craftsmanship which present the learning process in highly normative terms, pre-supposing a natural novice who gradually becomes a member of society by assimilating its superorganic heritage. Normative theory, as Lave (1988) points out, assumes a one-way, hierarchical ordering of knowledge:

> 'In this theory, duality of the person translates into a division of (intellectual) labour between academics and "the rest" that puts primitive, lower class, (school) children, female, and everyday thought in a single structural position *vis-à-vis* rational scientific thought.'

Normative theory necessarily dismisses everyday language as cheap and theoretically irrelevant, as 'mere' household words or 'loose talk'.

In discussions of fisheries management, contrary voices are also raised at times. Knowledge of the ecosystem, it is argued, is too imperfect for making reliable forecasts. Some scholars argue that multi-species fisheries are chaotic systems with too many uncertainties for any kind of control; such arguments have recently been developed in the scholarly literature on fisheries management (Wilson, 1982; Smith, 1990; Wilson & Kleban, 1992; Gilbertsen, 1993, Wilson *et al.*, 1994). If marine ecosystems are chaotic and fluctuating regimes, those who are directly involved in resource use are likely to have the most reliable information as to what goes on in the system at any particular point in time.

Practice theory offers a compelling view of learning, craftsmanship, and ethnography (Pálsson, 1994), a view which has important implications for management. Informed by the notions of situated action, mutual enskillment, and communities of practice, practice theory emphasizes both democratic communion and the continuity of the social world. The practice perspective resonates with some aspects of the folk discourse of Icelandic fishermen. In Iceland, experienced skippers often speak of knowing the details and the patterns of the 'landscape' of the sea bottom 'like the back of their hands'. This indicates that for the skilled skipper, fishing technology – the boat, electronic equipment, and fishing gear – is not to be regarded as an 'external' mediator between his person and the environment but rather as a bodily extension in quite a literal sense. Thanks to such technological extensions the experienced skipper is able to 'see' the fish, an otherwise invisible prey, and the landscape of the sea bed.

Moreover, 'real' schooling is supposed to take place in actual fishing. The emphasis on 'outdoor' learning is emphasized in frequent derogatory remarks about the 'academic' learning of people who have never 'peed in salty sea' (*migið í saltan sjó*). Questioned about the role of formal schooling, skippers often say that what takes place in the classroom is more or less futile as far as fishing skills and differential success are concerned, although they readily admit that schooling has some good points, preventing accidents and promoting proper responses in critical circumstances involving the safety of boat and crew. Even a novice fisherman, they say, with minimal experience of fishing, is likely to know more about the practicalities of fishing than the teachers of the Marine Academy. No formal training can cope with the flexibility and variability in the real world. Therefore, there is little, if any, connection between school performance and fishing success.

Skipper education recognizes the importance of situated learning. Earlier participation in fishing, as a deck hand (*háseti*), is a condition for formal training, built into the teaching programme; this is to ensure minimum knowledge about the practice of fishing. Once the student in the Marine Academy has finished his formal studies and received his certificate, he must work temporarily as an apprentice – a mate (*stýrimður*) – guided by a practising skipper, if he is to receive the full licence of skipperhood. The attitude to the mate varies from one skipper to another; as one skipper remarked, 'some skippers regard themselves as teachers trying to advise those who work with them, but others don't'. While skippers differ from one another and there is no formal economic recognition of their role in this respect, in terms of a teaching salary. According to many skippers the period of apprenticeship is a critical one. Reflecting on his mentor, with whom he had spent several years at sea, one skipper explained: 'I acquired my knowledge by working with this skipper, learning his way of fishing. I grew up with this man.' It is precisely here, in the role of an apprentice at sea, that the mate learns to attend to the environment as a skipper. Working as a mate under the guidance of an experienced skipper gives the novice the opportunity to develop attentiveness and self-confidence, and to establish skills at fishing and directing boat and crew. The role of the mate institutionalizes what Lave & Wenger (1991) term 'legitimate peripheral participation', a form of apprenticeship that allows for protection, experimentation, and varying degrees of skill and responsibility. This is not a one-way transfer of knowledge as the skipper frequently learns from the co-operation of his mate; mate and skipper educate each other. In the beginning, the mate is just like an ordinary deck hand; in the end he is knowledgeable enough to have a boat of his own. At first he is of little help to his tutor, later on he can be trusted with just about anything; occasionally, the skipper may even take a break and stay ashore, leaving the boat and the crew to his mate.

Often the advice of the skipper is in the form of verbal directions. He will draw the novice's attention to various aspects of skipperhood – how to manoeuvre the boat, how to use electronic equipment, how to follow fish migrations, and so on. A skipper is unlikely to share his most personal tricks, as the mate may later on become one of his competitors, in charge of another local boat. Every skipper has a personal 'diary' with details about times, species, catches, and fishing locations, and mates are rarely, if ever, given direct access to this information (some of it is stored in the memory of a computer located in the wheel house and, significantly, passwords are sometimes used for protection). On the other hand, a keen novice will gradually learn to imitate the actions of his skipper, observing his decisions and carrying out his commands. One skipper pointed out that he 'simply' learned most of what he knew 'by seeing how others handled their tasks'.

The skipper's knowledge is complex; a skipper must choose times and places to fish on the basis of a series of detailed environmental information. It is not surprising, therefore, that fishermen often refer to the importance of 'attentiveness' (*eftirtekt, athygli*) and 'perceptiveness' (*glöggskyggni*); the ability to recognize and apply an array of minute but relevant details. Attentiveness is a complex ability and includes, for example, being able to 'read' the sky and predict the weather, to participate in discussions within the local fleet, to understand the 'sparks' of electronic instruments, and to be able to co-ordinate crew activities. In order, however, to have success in catching fish, the skipper must dwell with his crew. Moreover, the crew is part of a larger context. Fishermen often speak of the personnel (*mannskapur*) of a boat in an extended sense, including several people ashore, those who ensure efficient repairs of equipment between fishing trips and those who bait lines and take care of nets, repairing old ones or supplying new ones. Indeed, folk accounts of fishing success often emphasize the importance of good fishing gear and the diligence of the people ashore responsible for its maintenance. 'Having a good crew', therefore, means not only being able to rely on a good fishing crew, but also being provided with good 'services' (*øjónusta*) on land.

The fleet is ever present as well. The fleets of different ports are hard to separate; during fishing they merge on the boundless sea. Nevertheless, the communion at sea is a very important one. Inevitably the skipper's decisions while fishing are constrained by the decisions of other skippers and by the movements of the fleet (see Durrenberger & Pálsson, 1986). While deciding where to fish is largely guided by the readings of electronic equipment and by the skipper's experience of earlier fishing seasons, of no less importance is knowing what other skippers are doing, where they are likely to be, and how much they will catch. There are obvious benefits in co-operating with other skippers and crews on a daily basis while at sea, especially after a long break in fishing; sharing information on the state of the major areas saves time and fuel and each skipper gains information about fish migrations that he could not acquire on his own. As Wilson (1990) points out, while fishermen seek to reduce the search problem by looking for recurrent patterns in the migration and location of fish, 'the number of observations necessary to establish regularity is far too large for any single individual to acquire.'

White (1989) proposes the notion of the 'fleet effect' to address this issue, suggesting that information sharing and variable discovery techniques within a fleet 'result in larger catches than if all boats fished alone'. However, because skippers compete among themselves for locations, fish, crew, and prestige, they often carefully guard valuable information available to them.

In some fisheries, for instance in Brazil (see Kottak, 1992), technological and economic changes seem to have resulted in rapid deskilling. There is little reason to believe that this has been the case in Iceland. While old and retired skippers sometimes point out that fishing has been radically transformed by electronic technology (including the computer), emphasizing that 'natural signs' are increasingly redundant, attentiveness continues to be one of the central assets of the good skipper and, just as before, it demands lengthy training. The skipper's universe is very different from that of his colleagues of earlier decades, but what shows on the screens of the radar, the computer, and the fish-finder is just as much a 'natural sign', directly sensed, as birds in the air or natural landmarks.

Given the complexity and uncertainty of marine ecosystems, fishermen's detailed monitoring of such systems, and their elaborate ecological knowledge, there may be good grounds for exploring more closely how fishermen acquire their practical ecological knowledge, how their knowledge differs from the textual knowledge of professional biologists, and to what extent the former could be brought more systematically into the process of resource management.

The social context: equity, distribution and ITQs

Studies of ITQ systems in fisheries and their effects are still in their infancy (for some examples in the emerging literature, see Dewees, 1989; McCay & Creed, 1990; 1993; Boyd & Dewees, 1992; Helgason, 1995). Changes in the actual distribution of ITQs as well as the direct and indirect responses to such changes represent an important field of future research. The following account of changes in the actual distribution of ITQs in Icelandic fishing is based on ethnographic interviews and descriptive statistics. Before proceeding, some general observations are in order (see also Durrenberger & Pálsson, 1987a; Árnason, 1993).

From the beginning of the Icelandic ITQ system, in 1984, boat owners could lease their quotas relatively freely. Permanent transactions with a vessel's ITQ share were, however, not permitted. Such transactions could take place indirectly when a vessel changed hands, but in general ITQs could only be transferred permanently between vessels if the 'donating' vessel was thereafter excluded from further participation in the quota system. With the new fisheries management legislation in 1990 several significant changes were made to the ITQ system. In particular, the former restrictions on permanent transactions with ITQs between boat owners were lifted – in other words, ITQs became fully divisible and independently transferable. Also, the ITQ system was expanded by including all 6–10 t fishing boats. These boats, not previously subject to effective catch limitation, were now allotted ITQ shares of the total allowable catch.

Because of the changes to the ITQ system with the 1990 legislation, analyses of quota distribution and its effects over time are somewhat problematic. It seems logical to analyse the period after 1990 in two different ways. In the first case, both the periods before and after are examined, excluding all 6–10 t boats with ITQs before and after

1990, thereby making comparison with the earlier period more acceptable. In the second case, analysis is exclusively limited to the latter period, including all boats with ITQs and observing to a greater extent the impact of the transferability of quotas, after the 1990 fisheries management legislation took effect. Note that while a boat owner is allotted ITQs in several species with different market values (cod, haddock, saithe, etc.), the overall size of each individual quota may be measured in terms of a basic unit or cod equivalent – an aggregate measure based on the relative market value of each species in terms of cod.

The distribution of ITQs: both periods (from 1984 to 1994)

Despite restrictions on direct transactions of ITQ shares before 1990, boat owners were still able to sell their ITQs indirectly. In particular, boat owners could sell their boats, and thereby their permanent share of the catch. As ITQs were attached to boats, boat owners could add to their share of the ITQ by buying other boats. If we look at total catch such indirect transactions of ITQs (see Fig. 5.1) – that is, vessel transactions – we see that they show a substantial increase from 1984 to 1990.

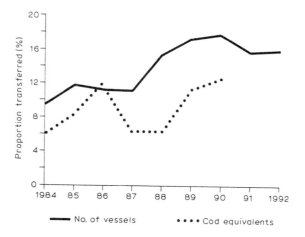

Fig. 5.1   Vessel and quota transactions, 1984–1994.

The proportion of ITQs, measured in cod equivalents, that changed hands through the sale of vessels doubled in six years, from 6% to 12.5% (the upswing in 1986 is due to the privatization of the fishing company that controlled the largest ITQ share, formerly the property of the capital city, Reykjavík). It is difficult to estimate the amount of capital involved in such transactions, but there were reports of vessels having been sold at a price two or even three times that of their 'real' value. Permanent access to the resource, therefore, was no less valuable in monetary terms than the vessel itself.

Boat owners were also able to increase their ITQ share by exploiting certain loopholes in the system. Prior to 1991, those who were dissatisfied with the size of their ITQ share could choose to fish under the effort quota system, in which case the size of the catch was less stringently restricted. If successful, a boat owner could then return his vessel to the proper system of ITQ after one year, with his boat's enlarged ITQ share. Boat owners could thus increase their boats' permanent share of the total allowable catch, at the expense of other boat owners, by fishing successfully under the system of effort quotas. Evidently, then, ITQs had been changing hands before the fisheries management legislation in 1990 lifted the restrictions on such transactions.

How were ITQs distributed and what changes in distribution have occurred over time? Fig. 5.2 shows changes in the total number of ITQ holders from 1984 (excluding 6–10 t boats for 1991–1994) and the relative size of four groups of ITQ holders, 'giants', 'large' owners, 'small' owners and 'dwarves' – those who own more than 1% of the ITQ shares, 0.3–1%, 0.1–0.3% and 0–0.1%, respectively. While being arbitrary, the demarcation of these groups provides a simple and effective way of discerning distributional changes of ITQs among boat owners, taking into account the fact that the distribution is positively skewed.

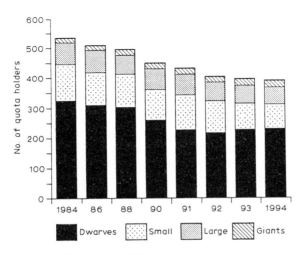

Fig. 5.2 Number of quota holders, 1984–1994.

As shown in Fig. 5.2., there is a constant and significant reduction (26.9%) in the total number of ITQ holders, from 535 to 391. Moreover, the number of 'giants' increases gradually from 1984, while all other groups generally diminish in number.

Fig. 5.3 shows the sizes of ITQ holdings belonging to the four groups of quota owners. Over the 11 year period the 'giants' increased their aggregate share by 78.4%, with their average share going from 1.64% in 1984 to 1.91% in 1994. The aggregate shares of all other groups decreased. Moreover, in the case of 'dwarves' and 'large' owners, average shares are slightly reduced – 0.039% to 0.038% and 0.58% to 0.52%, respectively. Consequently, in addition to the concentration caused by the decrease in the number of ITQ holders, the larger companies seem to be accumulating a disproportionate share of the quotas.

It is evident that while ITQs were gradually becoming concentrated in the hands of fewer companies up to 1990, the process of concentration gained increased momentum after 1990 when the new legislation made quotas directly transferable. This is perhaps inevitable, given that the main purpose of transferability of ITQs was to achieve efficiency in the fishing industry, making it easier for profitable fishing companies to buy out the quota of those that were less well managed.

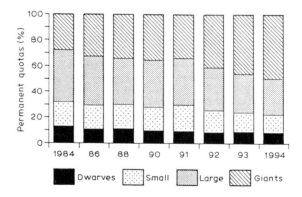

Fig. 5.3 Distribution of quotas (%) by size of holder, 1984–1994.

The distribution of quotas: the latter period (1991–1994)

With the fisheries management legislation passed in 1990, 704 new ITQ holders were added to the ITQ system in the demersal fisheries. Bearing in mind the aforementioned objectives of efficiency to be achieved by the transferability of ITQs, what changes in distribution of ITQs have occurred since 1991? As before, one way of establishing this is to examine changes in the number of ITQ owners (including owners of 6–10 t boats). Fig. 5.4 shows changes in the total number of ITQ holders and the relative size of the four groups of ITQ holders defined above. First, there is a constant and significant reduction (26%) in the total number of ITQ holders, from 1155 to 855. Moreover, while the number of 'giants' increases by 62.5%, all other groups diminish in number – the most notable case being that of the 'dwarves' whose number decreases by 254 in four years (26.7%). These figures plainly indicate the growing concentration of ITQs in the hands of the larger companies, which seem to be buying out a significant proportion of the smaller ITQ holders. It is especially interesting to note that most of the ITQ holders categorized as 'dwarves' were newcomers to the ITQ system in 1991 – the owners of 6–10 t boats.

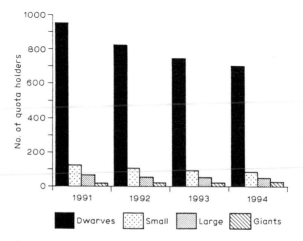

Fig. 5.4 Number of quota holders, according to size of holder, 1991–1994.

The reduction in the number of ITQ holders indicates an increased concentration of ITQs among those that remain. Fig. 5.5 shows that the 'giants' increase their permanent share of the total allowable catch by 85% (from 25.54% to 47.23%), with their average share going from 1.60% in 1991 to 1.82% in 1994, despite being the only group to increase in number. The aggregate share of all other groups diminishes. However, due to a severe decrease in numbers, the average ITQ shares of 'dwarves' and 'small' ITQ holders increase slightly; 0.018% to 0.019% and 0.16% to 0.17%, respectively. Thus, an analysis of the latter period, including all ITQ holders, reveals that ITQ shares are becoming increasingly concentrated in the hands of fewer holders, especially at the top.

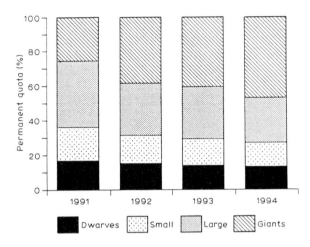

Fig. 5.5   Distribution of quotas (%) by size of holder, 1991–1994.

We need to keep in mind, though, that some, if not all, of the 'giant' companies are owned by a large number of share holders. One could argue, therefore, that the concentration of ITQs described above really masks a more egalitarian distribution of access and ownership. However, this is not necessarily the case. For one thing, it is quite probable that some individuals own shares in several different companies; thus, the distribution of ownership of ITQs may be even more unequal than our figures indicate. Moreover, the distribution of holdings within the biggest companies' shareholders is generally very positively skewed, with few controlling the majority of the shares – making the total number of share holders insignificant. Finally, even though share holders turned out to be more numerous than before, in actual practice a small group of managers have immense powers in their hands; they are the ones who control access to the resource, how the resource is used, what happens to the products, and how the benefits are distributed. These managers effectively control the fate of whole communities in Iceland that increasingly depend on one or two large ITQ holding companies for employment and economic existence in general – and this dependency has grown more acute with the ongoing concentration of fishing rights.

Equity, efficiency and 'quota profiteering'

Following bleak estimates of the fish stocks by marine biologists, politicians have made recurrent cuts to the total allowable catch since 1989, effectively devaluing ITQ shares. Consequently, many small companies were left with insufficient quotas to keep their boats active throughout the fishing season. While a number of small companies seem to be persevering, many were forced to sell their ITQs and leave the ITQ system. Meanwhile, other more affluent companies, often referred to in public discourse as 'quota kings' (*kvótakóngar*) and 'lords of the sea' (*sægreifar*), have been accumulating fishing quotas, in some cases more than they are capable of using themselves.

One interpretation of the data presented above is that the objectives of efficiency in the fishing industry are being attained. Hence, it may be argued, the 'invisible hand' of the market has simply shifted the distribution of ITQs to a state of greater efficiency, by reducing the number of ITQ holders and concentrating fishing rights in the hands of larger companies. However, as the results presented above indicate, if there is an increase in 'efficiency' it only takes place at the cost of equity. Indeed, many scholars have suggested that such a trade-off is inevitable (Okun, 1975; Gatewood, 1993). With some companies holding more than they are capable or willing to fish and others with less than they actually need, ITQs seem to be 'inefficiently' distributed. A characteristic market solution to such a problem, perhaps, and one that is increasingly prevalent in the Icelandic ITQ system, is that of a leasing. A company will temporarily rent a part of its ITQ share to a company lacking ITQ; in public discourse this is frequently referred to by the loaded term of 'ITQ profiteering' (*kvótabrask*).

The standard lease price paid by lessees in such transactions is approximately 40% of the value of the catch. For the lessor companies, then, 'quota profiteering' represents a lucrative business transaction. By leasing a part of its share, a company can free itself from the expenses of actually catching the fish, while still procuring 40% of the value of the resulting catch. Fig. 5.6 outlines the *net* traffic of leased ITQs between the four groups defined above for a three year period. In 1991 the main 'donors' of leased ITQs

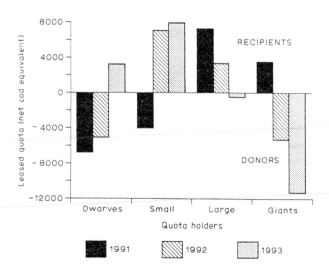

Fig. 5.6 Net movement of quotas by size of holder, 1991–1993.

were the 'dwarves' and 'small' ITQ holders; 'large' ITQ holders and 'giants', on the other hand, were recipients. This situation may largely be explained by the fact that many of the new small-scale operators found their ITQ allotments insufficient to maintain their fishing enterprise and, as a consequence, they placed their fishing rights on the market. At the same time, the larger companies were attempting to increase their share in the total allowable catch. However, in the following two years, concurrent with a substantial increase in the concentration of ITQs in the largest companies, the traffic of leased ITQs shifted dramatically. Thus, in 1993 the 'giants' are almost the sole 'donors' of leased ITQs while 'dwarves' and 'small' ITQ holders are now 'recipients'.

This state of affairs has led many fishermen to describe the ITQ system in feudal terms, with the 'quota kings' or 'lords of the sea' controlling most of the quotas and profiting from renting it to 'tenant' companies, who actually do much of the fishing. After paying the lease price, the 'tenant' companies are left with only 60% of the value of the catch, while still bearing all the normal expenses of fishing. In the 'tenancy' system it is the 'quota kings' who are in charge; not only do they own most of the ITQs, they also control many of the plants that buy the catch. Thus, to quote one skipper: 'one must give in to almost every demand, because the quota king makes all the rules, sets the price and everything'. However, the 'quota kings' themselves view the matter from a totally different perspective, maintaining that most of the so-called 'lords of the sea' are really on the verge of bankruptcy, citing envy as the main source of the feudal metaphors.

By law, fishermen receive a fixed share of the value of the catch. Understandably, however, the 'tenant' companies try to minimize their additional costs, represented by the leasing of ITQs, by cutting the wages of fishermen. Increasingly they reckon fishermen's wages from the amount left after the lease price has been subtracted from the value of the catch. In such cases, fishermen working for 'tenant' companies may suffer up to 40% wage cuts – in effect 'paying' a sizeable part of the lease price of ITQs to the 'quota king'. Significantly, Icelandic fishermen went on a national strike in January 1994, protesting against 'quota profiteering' and the 'tenancy' system, leading to a two week stand-still in the fishing industry. The Government put an end to the strike by passing temporary laws to force the fishermen back to work. In May 1995 Icelandic fishermen went on strike again. This time the strike lasted for three weeks, ending with an agreement involving concessions on both sides. While the feudal metaphors used in public discourse are, perhaps, not wholly warranted, it seems that the concentration of fishing rights in the hands of the largest companies is having a far reaching effect on the Icelandic fisheries and, more generally, Icelandic society.

Final remarks

There is a strange paradox in Western environmental discourse. On the one hand, we tend to project a modernist image of resource management as an apolitical enterprise, as the 'rational' domination of nature by means of mathematical equations, independent of ethics and social discourse. On the other hand, modern environmental discourse is characterized by the postmodern condition, emphasizing, much like medieval European discourse, the embedded nature of any kind of scholarship and the interrelatedness of nature and society (Merchant, 1980; Gurevich, 1992). The former view, which presents the pursuit of environmental knowledge as a relatively straightforward accumulation of 'facts' and radically separates knowledge of nature and the social context in which it is produced, has come increasingly under attack in several fields of scholarship, including anthropology, economics, and environmental history. Some scientific communities, however, stick to Baconian notions of scientific methods, of 'observation', and the domination of nature. Policy makers in fisheries often remain firmly committed to a modernist stance, curiously innocent of recent developments in social and ecological theory, presenting themselves as detached observers, as pure analysts of the economic and material world, independent of the 'partial' viewpoints and the trivial, practical knowledge of the actors (see Durrenberger, 1992). We may well be advised to search for alternative epistemologies and management schemes, democratizing and decentralizing the policy making process. It may simply be more effective.

One illustration of the modernist paradigm with respect to resource management is the suppression of inequality and social distribution, a theme frequently pushed to the margin (for a different view, see Sen, 1973). The analysis presented above of the Icelandic cod fishery, focusing on changes in the actual distribution of ITQs, indicates a growing inequality; fishing rights have been increasingly concentrated in the hands of the biggest companies. Also, with effective transferability of ITQs after 1990, the concentration of ITQs has escalated. In 1994, only 26 companies (the 'giants') own about half of the national quotas in the demersal fisheries. Significantly, public discontent with the concentration of fishing rights and its social and political repercussions, including 'quota profiteering', is increasingly articulated in terms of heavily loaded feudal metaphors, of 'tenancy', 'quota kings', and the 'lords of the sea'. The issues of equity, distribution, and social discourse should not be seen as trivial externalities devoid of practical and theoretical significance. If the players perceive the system as unjust, they are likely to find some ways of cheating, of bending or avoiding the rules, and that may make it difficult to achieve the stated goals of management. Consequently, comparative studies by social scientists of the unexpected as well as the foreseen consequences of management regimes represent an important topic for research.

Acknowledgements

The study on which this article is based is part of two larger, collaborative research projects, 'Common Property and Environmental Policy in Comparative Perspective', initiated by the Nordic Environmental Research Programme 1993–1997 (NERP), and 'Property Rights and the Performance of Natural Systems', co-ordinated by Susan Hanna and funded by the Beijer Institute of the Swedish Academy of Sciences. Work on this article has been supported by several other programmes and institutions, including the Nordic Committee for Social Science Research (NOS-S), the University of Iceland, and the Icelandic Science Foundation. The Icelandic Ministry of Fisheries kindly provided much of the raw statistical data used. The empirical parts of the article focusing on the distribution of quotas were originally presented to a workshop on 'Building resilience, equity, and stewardship into market approaches to the resolution of common property problems', Rutgers University, 9–12 March 1994.

References

Acheson, J.M. (1988) *The Lobster Gangs of Maine.* Hanover and London University Press, New England.

Arnason R. (1993) Icelandic fisheries management. In *The Use of Individual Quotas in Fisheries Management*, pp. 123–43. OECD, Paris.

Boyd, R.O. & Dewees, C.M. (1992) Putting theory into practice: transferable quotas in New Zealand's fisheries. *Society and Natural Resources*, **5**, 179–98.

Dewees, C. M. (1989) Assessment of the implementation of individual transferable quotas in New Zealand's inshore fishery. *North American Journal of Fisheries Management*, **9**, 131–9.

Dilley, R. (ed.) (1992) *Contesting Markets: Analyses of Ideology, Discourse and Practice*. Edinburgh University Press.

Durrenberger, E.P. (1992) *It's All Politics: South Alabama's Seafood Industry*. University of Illinois Press, Chicago.

Durrenberger, E.P. & Pálsson, G. (1986) Finding fish: the tactics of Icelandic fishermen. *American Ethnologist*, **13**, 213–29.

Durrenberger, E.P. & Pálsson, G. (1987a) The grass roots and the state: resource management in Icelandic fishing. In *The Question of the Commons: the Culture and Ecology of Communal Resources* (eds B.M. McCay & J.M. Acheson), pp. 370–92. University of Arizona Press, Tucson.

Durrenberger, E. P. & Pálsson, G. (1987b) Ownership at sea: fishing territories and access to sea resources. *American Ethnologist*, **14** (3), 508–22.

Ferber, M.A. & Nelson, J.A. (Eds) (1993) *Beyond Economic Man: Feminist Theory and Economics*. University of Chicago Press, Chicago.

Gatewood, J.B. (1993) Ecology, efficiency, equity, and competitiveness. In *Competitiveness and American Society*, (ed. S.L. Goldman), pp. 123–55. Lehigh University Press, Betlehem, PA.

Gilbertsen, N. (1993) Chaos in the commons: salmon and such. *Maritime Anthropological Studies*, **6** (1/2), 74–91.

Gudeman, S. (1986) *Economics as Culture: Models and Metaphors of Livelihood*. Routledge & Kegan Paul, London.

Gudeman, S. (1992) Remodeling the house of economics: culture and innovation. *American Ethnologist*, **19** (1), 139–52.

Gurevich, A. (1992) *Historical Anthropology of the Middle Ages*, (ed. J. Howlett), Polity Press, Oxford.

Hanna, S.S. (1990) The eighteenth century English commons: a model for ocean management. *Ocean and Shoreline Management*, **14**, 155–72.

Helgasson, A. (1995) *The Lords of the Sea and the Morality of Exchange: the Social Context of ITQ Management in the Icelandic Fisheries*. M.A. thesis, University of Iceland.

Kottak, C.P. (1992) *Assault on Paradise: Social Change in a Brazilian Village*. 2nd edn. McGraw-Hill, New York.

Lave, J. (1988) *Cognition in Practice: Mind, Mathematics and Culture in Everyday Life*. Cambridge University Press, Cambridge.

Lave, J. & Wenger, E. (1991) *Situated Learning: Legitimate Peripheral Participation*. Cambridge University Press, Cambridge.

McCay, B.M. & Acheson, J.M. (eds) (1987) *The Question of the Commons: the Culture and Ecology of Communal Resources*. University of Arizona Press, Tucson.

McCay, B.M. & Creed, C.F. (1990) Social structure and debates on fisheries management in the Atlantic surf clam fishery. *Ocean and Shoreline Management*, **13**, 199–229.

McCay, B.M. & Creed, C.F. (1993) *Social Impact of ITQs in the Sea Clam Fisheries.* Rutgers University, Mimeo.

McCloskey, D. (1985) *The Rhetoric of Economics.* University of Wisconsin Press, Madison, WI.

McEvoy, A.F. (1988) Toward an interactive theory of nature and culture: ecology, production, and cognition in the California fishing industry. In *The Ends of the Earth: Perspectives on Modern Environmental History* (ed. D. Worster), pp. 211–229. Cambridge University Press, Cambridge.

Merchant, C. (1980) *The Death of Nature: Women, Ecology and the Scientific Revolution.* Harper & Row, San Francisco.

Neher, P.A., Árnason, R. & Mollett, N. (eds) (1989) *Rights Based Fishing.* Kluwer Academic Publishers, Dordrecht.

Okun, A.M. (1975) *Equality and Efficiency: the Big Trade-off.* The Brookings Institute, Washington, DC.

Pálsson, G. (1991) *Coastal Economies, Cultural Accounts: Human Ecology and Icelandic Discourse.* Manchester University Press, Manchester.

Pálsson, G. (1994). Enskilment at sea. *Man,* **29** (4), 901–27.

Pálsson, G. & Durrenberger, E.P. (1990) Systems of production and social discourse: the skipper effect revisited. *American Anthropologist,* **92**, 130–41.

Roseberry, W. (1991) Potatoes, sacks, and enclosures in Early Modern England. In *Golden Ages, Dark Ages: Imagining the Past in Anthropology and History.* University of California Press, Berkeley.

Scott, A.D. (1989) Conceptual origins of rights based fishing. In *Rights Based Fishing* (ed. P.A. Neher, R. Árnason & N. Mollett), pp. 11–38. Kluwer Academic Publishers, Dordrecht.

Sen, A. (1973) *On Economic Inequality.* Oxford University Press, Delhi.

Smith, M.E. (1990) Chaos in fisheries management. *Maritime Anthropological Studies,* **3** (2), 1–13.

Vestergaard, T. (1993) Catch regulation and Danish fisheries culture. *North Atlantic Studies,* **3** (2), 25–31.

White, D.R.M. (1989) Knocking 'em dead: Alabama shrimp boats and the 'fleet effect'. In *Marine Resource Utilization: a Conference on Social Science Issues* (eds J.S. Thomas, L. Maril, & E.P. Durrenberger), University of South Alabama College of Arts and Sciences Publication Vol. 1 and the Mississippi-Alabama Sea Grant Consortium, Mobile.

Williams, R. (1976) *Keywords: a Vocabulary of Culture and Society.* Fontana, Glasgow.

Wilson, J.A. (1982) The economical management of multispecies fisheries. *Land Economics,* **58** (4), 417–34.

Wilson, J.A. (1990) Fishing for knowledge. *Land Economics,* **66**, 12–29.

Wilson, J.A., Acheson, J.M., Metcalfe, M. & Kleban, P. (1994) Chaos, complexity and community management of fisheries. *Marine Policy,* **18** (4), 291–305.

Wilson, J.A. & Kleban, P. (1992) Practical implication of chaos in fisheries. *Maritime Anthropological Studies,* **5** (1), 67–75.

Chapter 6
Trade in Fishing Rights in the Netherlands: a Maritime Environment Market

Ellen Hoefnagel

Introduction

Just occasionally, a phenomenon occurs in reality that theorists only dare dream about. Such is the case with the development of ITQs in the Netherlands which provides an opportunity for analysing the actual processes associated with a market orientated environmental policy. This chapter, based on interviews with fishermen and board members of fishermen's organizations in the Netherlands in 1992 and 1993, examines the spontaneous development and operation of a quota market for sole and plaice. This trade in fishing rights simulates the system of tradable licences often proposed as an instrument for reducing the level of environmental pollution.

Above a certain level of pollution, the quality of the environment becomes threatened. Where a maximum acceptable level of pollution can be stipulated, this level can be divided into 'polluting rights', whereby the polluter may only pollute to the level of the rights expressed in a pollution permit or licence. Licences to pollute represent an economic value; pollution costs money. Entrepreneurs will therefore draw up calculations of if and when pollution is still cost efficient. At the same time, the costs of pollution will help to bring about a great incentive for developing and applying a more environmentally friendly technology. Pollution rights should be negotiable and economically inefficient firms may consider cashing in their pollution rights through trading them on a licence market. Although, in environmental economics, a licence market is considered mainly in relation to pollution, here the trading of fishing rights is compared to a licence market. Pollution involves the discharge of damaging substances into the environment; but it can prove equally harmful to abstract certain substances from the environment beyond a given level. Where, for example, water is scarce, high levels of abstraction can cause a deterioration in water quality and create drought.

Overfishing can deplete the stocks of fish to a point where the ability to replenish the resource is seriously harmed. To prevent such environmental damage, inventory control systems are brought into play to protect the fish stocks. This occurs in the fishing grounds of the EU as a whole, where maximum quantities are set for catches (total allowable catches or TACs). In principle, the question of whether the environment can be damaged by the addition or subtraction of substances does not materially affect discussion of the environmental effects of a licence market. In each case a ceiling is set; on the one hand, it limits the amount of pollutants that may be discharged into the environment and, on the other, the amount of fish that may be abstracted from the environment.

Trade in pollution rights has not yet begun in the Netherlands. The economist Nentjes has long argued for negotiable pollution rights, which he terms 'environment use rights' (Nentjes, 1988). So far the operation of negotiable environment use rights has only been theorized. Kunnecke (1992) first associated the idea of catch quotas with the environmental economic theory that negative externalities of production (including pollution) should be held in check by market mechanisms rather than by government regulation. Kunnecke (1992) argues that certain government regulated markets, where quotas are already in place, are suitable for the application of environmental economic theories. However, a prior condition is that quotas become individualized and negotiable. Fishing quotas for sole and plaice have been individualized and transferable since 1977 and officially negotiable since 1985. This, therefore, offers an excellent opportunity for investigating how the trade in environmental use rights operates in practice.

This contribution thus investigates both the social and economic effects of the trade in fishing rights for the fisherman-entrepreneur and also the effects on the environment, that is upon the stocks of sole and plaice. A broader question is also implied, namely the extent to which such a market can contribute to environmental controls and perhaps replace government regulation.

The establishment of ITQs

Since 1975, TACs have been defined in the North Atlantic for each major commercial fish stock and for each fishing area. The size of the TACs is reset annually on the basis of biological advice prepared by the International Council for the Exploration of the Sea (ICES) and since 1977 the political decision making has been undertaken by the Council of Fisheries Ministers. The limits are therefore flexible. The EU receives its TACs and divides these among the member states according to the formula agreed in 1982 as part of the Common Fisheries Policy (CFP). Thus each Atlantic coast member state receives a given share of the potential 'Eurosea' harvest. The Dutch quota for sole amounts to 75% of the North Sea sole TAC and for plaice the national quota represents 38% of the North Sea TAC. Member states have the freedom to stipulate how much and at what rate their allocation can be fished, as long as the national quota is not exceeded. When the national quota has been exhausted, no more fishing for that species is allowed.

Before creating a licence market, a number of basic arrangements must be agreed: who has the rights; the date for commencement of trading; and the rules of negotiability (Peeters, 1992). A right or licence is a consent from the authorities to prosecute an action which would otherwise be illegal. It must therefore be clear precisely what is to be permitted and the period of validity of the right must be established. The licence market must be initiated by the Government through a division of rights among the applicants by way of sale or by allocating the rights free of charge on the basis of administrative criteria.

In the Netherlands, the national quotas for sole and plaice were divided among individual fishing enterprises on the basis of historic catch performance and the catch capacity of the vessel, based on engine size. This was known as the 'grandfathering' system (Peeters, 1992; Bressers, 1985). For each vessel issued with sole and plaice licences, a maximum quantity of catch is specified for a given time period. Formerly, the annual rules were set down in 'current decrees for catch limit settlements' with additional specifications for the year in question. The CFP, under which the TACs for sole and plaice are agreed, was set for successive periods: 1977–1982; 1983–1992; and currently 1993–2002. The duration of the fishing right, therefore, was for a minimum of one year but with an expectation of extension for at least ten years; the rights thus entail a degree of uncertainty about their continuation in the longer term.

Negotiability of the fishing right means that it is possible to transfer the quota from one fisherman to another, as for example from father to son. On closing down the enterprise, a fisherman may also transfer his quota to another fisherman, providing the recipient already holds an appropriate engine capacity licence and is in possession of a sole and plaice licence. In the event of a fisherman being unable to fulfil his quota, for example, due to his boat being laid up for part of the year, he is not permitted to make up this lost quota in the following year. He can, however, transfer the unfished share of the quota to a colleague. Fishermen view their fishing rights as a personal possession; when a fisherman decides to wind up his enterprise, he not only offers his vessel for sale but also his fishing rights. Fishing rights have become negotiable and fishermen have thus spontaneously translated environmental economic theory into practice. Such transactions must be approved and registered by the Ministry in order to maintain central control of the fishery (Salz, 1991).

In summary, it can be claimed that the transfer of fishing rights satisfies a number of core elements of a licence market. Only fishermen with specific entitlements to sole and plaice have the right to fish for these species. The division of rights is accomplished through the 'grandfathering' principle. The conditions for limitation of the catch are clearly set down by means of annual TACs for specific species and fishing areas. The rights are transferable. Duration of the right is, in the short term, assured but in the longer term uncertain. The role of the Government is clear: through the Council of Ministers, it agrees the TACs on the basis of advice from ICES. It is, however, worth noting that the introduction of the 'grandfathering' and the commencement of trading in fishing rights are not coincident and that the Dutch Government keeps itself rather aloof from the trade. It may even be doubted that the negotiability of fishing rights was ever intended.

> 'Look, the Ministry said you may transfer your quota. But we ourselves put a price on them'.
>
> (A Dutch fishermen)

Development of a market for fishing rights

In the period 1975–86 a 'grey market' for fish developed in the Netherlands. A detailed analysis of the origins of the 'grey market' falls outside the aims of this chapter. One reason was the doubt shared by a number of Dutch fishermen over the legitimacy and legality of the EU's authority. A CFP, endorsed by all member states, came into force in 1983. A second reason was that enforcement of quota regulations left something to be desired. Since 1983 member states were required to implement the fishery regulations. Yet the period from 1983 up to and including 1986 was, in many respects, comparable with the previous period. Up to 1986 individual quotas were to a large degree exceeded: too many fish were caught and earnings were high. Partly with the aid of government subsidies, fishermen invested in new, modern vessels with greater engine capacity. There was talk of business expansion and new entrants, mostly ex-crew members, also procured vessels.

An evaluation of fisheries regulations by the EU in 1986 concluded that in many member states there was a serious shortfall in the standards of supervision and even talk of opposition from national administrations. According to the evaluation, the Netherlands scored worst among the member states. Questions were asked in Parliament, which resulted in an inquiry in 1987 into the active and passive involvement of the Dutch Ministry with regard to the evasion of EU quota regulations. Although these investigations had few political consequences, they did have important implications for the fishing industry which was to become much more strictly controlled concerning the observance of quota regulations. Each fisherman was only allowed to fish for sole and plaice in accordance with his own individual fishing rights. For some fishermen, it became increasingly important to obtain additional rights over and above the level of their historic track records.

A few fishermen were also anticipating the suppression of the 'grey market'. One such had up to 1984 not invested in a new vessel. When he finally ordered a vessel to be built for delivery in 1984, the 'grey market' was rumoured to be dying out. His existing fishing rights could be transferred from the old vessel to the new one. But as his old vessel had almost no historic rights attached to it, he faced a potential problem in paying for the new vessel when the catch could no longer be landed on the 'grey market'. He decided to buy a quota. For approximately 200 000 guilders he purchased a quota for sole and plaice – a daring step, as it meant a considerable investment on top of that for the replacement of the boat. He paid 7.0 guilders per kilo for fishing rights for sole. Three months later, his brother-in-law paid 15.0 guilders per kilo. Since then the price of fishing rights has risen steadily; by 1992, 114.0 guilders per kilo was being asked for fishing rights for sole.

In 1990 approximately 900 tonnes of sole fishing rights and 4300 t of plaice fishing rights were traded (Ministerie van LNV, 1992). The national quota for sole totalled 18 750 t in 1990; for plaice the national quota was 68 400 t. Thus roughly 5% of all fishing rights for sole and 6% for plaice were traded in 1990.

The trade in fishing rights

Logically, the price of fishing rights in a balanced market should react to changes in the annual level of TACs: the consequences of a decrease in the TAC should be an increase in value of the quota. The fishing rights became scarcer and the value of the landed fish also increases on the market. As can be seen in Table 6.1, TACs for sole were comparatively low from 1987 to 1989 but were subsequently greatly increased. Plaice, meanwhile, had a low TAC in 1987 but in other years remained fairly stable. Since 1986, however, there has been an enormous increase in the value of the quotas. 'The price levels specifically became set through a few capital rich fishermen.' The financing of the purchase price comes from their own assets or through the increase of the mortgages on their ships. As well as the uncertainty of the industrial branch over the long term, banks do not recognize the value of the fishing quota when increasing loans (De Boer & Van Keulen, 1992). It is difficult for a firm with a high mortgage burden to invest in fishing rights. There is, therefore, a tendency for strong businesses to grow stronger and weak businesses weaker. In one Dutch fishing port, an estimate of the distribution of quotas made by a fisherman suggested that of the 60 vessels, 16 had approximately 40–50% of the quota (Hoefnagel, 1993).

Table 6.1   Agreed TACs for sole and plaice, 1984–1991 (thousand tonnes).

	1984	1985	1986	1987	1988	1989	1990	1991
Sole	20	22	20	14	14	14	25	27
Plaice	182	200	180	150	175	185	180	175

Source: ICES, 1992.

In 1993 a change occurred. The high prices for fishing rights of 114 guilders/kilo for sole and 14.5 guilders/kilo for plaice fell significantly in 1992 due, among other things, to an increase of 88% in the sole TAC. True, the TAC for plaice declined but the price per kilo of plaice quota also fell as it is closely bound up with the price for sole quota and also because the market for boneless plaice had declined. And, according to the fishermen, a further reason was the reduced stocks of plaice so that the fishermen were experiencing difficulty in fulfilling their quotas.

Fishing enterprises with few rights also have the opportunity to lease them. A fisherman with surplus quota can temporarily rent out his fishing rights. In 1990, 600 rental transactions were completed (Ministerie van LNV, 1992) involving 553 vessels. This represents an extensive market. According to the Ministry, rental prices in 1990 were 6 guilders per kilo for sole and 1 guilder per kilo for plaice.

Sole and plaice fishermen thus sell or rent fishing rights to each other. In the Netherlands there are also fishermen who have acquired fishing rights in the UK and Germany. They came into possession of British sole and plaice licences at relatively low cost as a result of their vessels 'switching flags'. In other words, Dutch vessels were sold to English-based fishing companies which were in fact controlled by Dutch fishermen. There are some 60 Dutch vessels fishing under the British flag that legally exploit fishing rights for sole and plaice assigned by the EU to the UK. Their catches are handled in the Netherlands and the vessels retain their original ports as their home base. According to Peeters (1992) 'in principle, the large application area of the EEC makes a common licence market for non-location specific environmental taxation activities attractive'. In Peeters' case, 'location specific' refers, for example, to a drainpipe emptying into a river. The EU's fishermen have an immense 'location' in which to make use of their fishing rights, namely the North Sea and neighbouring areas of the Atlantic. Dutch fishermen have, in a sense, created a common licence market. Through 'flagship' practices, the relationship between fishing capacity and fishing opportunities in the UK fleet is coming under pressure. Comparable transactions were also undertaken in Germany but constraints have also been introduced. By contrast, it has been virtually impossible for foreign fishermen to obtain Dutch fishing rights because of the prohibitively high prices. This spontaneous and insufficiently regulated common licence market also has its darker side because of unequal competition among fishermen in different member states.

Government, inland revenue and the fishing rights trade

The value of fishing rights is set on a free market. At the start of quota setting in 1975, there was a proposal to structure the market by means of a quota management fund. In the case of a firm ceasing to fish, the quota would be surrendered to a fund managed and financed jointly by the Government and the industry. After surrender, the quota was to be evenly divided among fishermen with existing sole and plaice fishing rights. Thus, enlargement of an individual's quota would be achieved passively. But, in practice, a 'wait and see' attitude does not suit entrepreneurs. Moreover, as the buy out premium for fishing rights was unlikely to be high, there would be little incentive for an inefficient enterprise to close. The quota management fund did exist briefly as part of a recognition scheme at the beginning of the quota setting process but, with the existence of a 'grey market', there was no real reason for a fisherman to make use of the reorganization scheme. In any case, after the introduction of stricter quota controls, the fishermen were willing to pay more through the free market to obtain quotas from firms ceasing to fish than a state-run buy-out fund could afford.

Strict conditions were imposed by the Government on the transfer of fishing rights. Rights to fish sole and plaice can only be traded by owners of vessels that are effectively listed on a central register and which possess appropriate sole and plaice licences and engine capacity licences. Furthermore, fishermen may not sell their fishing rights in portions, only in their entirety – although purchasers can buy portions. This may seem contradictory, but on the market the quota can only be sold if enough fishermen can be found to buy the entire quota outright. The transfer must be approved and registered by the Ministry who also require the buyer to provide a procurement consent from the bank.

The purchase of fishing rights is tax deductible, with the effect that profit-making enterprises frequently buy up relatively small portions of fishing rights in order to reduce their tax obligations and thereby drive up the purchase price. The opportunity to write off quota purchases is itself remarkable in the sense that this normally applies only to goods that will eventually wear out. Indeed, the regional tax office in Breda for a time refused to consider quota write offs while tax inspectors in Zwolle and Zeeland were willing to allow them, thus giving rise to a discrepancy within the system. In 1993, however, opportunity was standardized for all regions.

There is a reverse side to this tax bonus. When a vessel is sold tax must be paid on the current value of the vessel, together with its gear and fishing rights. This also occurs in the case of succession within a family business. A transfer made against the opinion of the inland revenue that the tariff was too low, can lead to payment on extra duty. In the case of milk quotas, succession rights and tax liabilities are negotiated on the basis of the value of the land area leased, set at one fifth of current market value. The successor wishing to take over the business from his father must pay tax on this notional value rather than the actual current value of the land plus the milk quota. But when fishing quotas are being transferred, current value is calculated as the base for the tax liability. In a stressed market, such as that for fishing rights, this can lead to problems: the quotas are highly valuable but there is no money available to pay for them. Fathers who fish have at some time received the fishing rights free of charge but their successors, meanwhile, can only with great difficulty take over these valuable assets because of the effect of succession rules and liabilities.

Social, economic and environmental effects of negotiable fishing rights

Fishing rights for sole and plaice have undergone increases in market value with certain significant socio-economic effects including increased difficulty of entry into the sector for newcomers; difficulties in business expansion; closure of inefficient and, in some instances, healthy enterprises; the concentration of fishing rights in larger enterprises; and the scaling down of enterprises – each with its own environmental effect.

Difficulty of entry into the sector by newcomers

Between 1977 and 1986 it was still possible for new entrants to start a fishing enter-prise, usually with only a small initial quota. After the increase in the level of control over the relationship between fishing rights and landed catches, this became virtually impossible. In 1977 there were 495 vessels (excluding shellfish boats and distant-water trawlers) with a total of 352k HP; ten years later, as a result of new entrants and the expansion of existing businesses, the figures had risen to 611 vessels and 581k HP (Ministeries van LNV, 1992). Since 1987 the number of vessels and aggregate engine capacity has fallen to 494 and 502k HP, respectively. As a result fishing capacity and fishing opportunities are no longer growing further apart, with clear environmental benefits in terms of pressure on stocks.

Difficulties of business expansion

The Dutch fishing fleet remains mainly in the hands of family enterprises. The crew of a vessel is ideally recruited from the neighbouring family, with the most common form comprising fisherman owner and his sons. When the owner becomes older and the sons are full grown – or already have sons of their own who want to fish – the family firm is divided between the sons. The father may continue to participate in the enterprise but to a diminishing degree or 'retire' to work onshore. His sons continue on their own vessels fishing with their sons, possibly supplemented by other crew members. This system is widespread throughout the North Atlantic fisheries and is known as the 'all brothers crew life cycle' (Jorion, 1982). But the division of the firm among brothers may no longer be practicable, due to the high value ascribed to the fishing rights. A solution to this impasse is still possible through the reflagging of vessels but this is difficult to do. Consequently, brothers must often remain fishing together on the original vessel. Only when there are two brothers, two vessels and a large bundle of fishing rights is the traditional division of the firm feasible. The disruption of the cycle of succession also implies a curtailment of expansion of the fleet, with similar beneficial environmental effects.

The closure of fishing firms and the concentration of fishing effort

Fishermen with too little quota can buy or rent additional quota. However, their prices are so high that it may no longer be cost-effective to acquire additional quota. Illegal 'black fish' landings have provided one 'solution' but such illegal activity is increasingly identified and punished. The closure of the small fishing enterprise then becomes the most likely solution. The quotas of outgoing firms are usually bought up by stronger, surviving profitable firms with sufficient quota and manageable mortgage obligations. This tendency for strong firms to become stronger is in line with the expected results of negotiable environment use rights (Nentjes, 1988; Kunnecke, 1992).

With the rising value of fishing rights, retirement and succession within the family business has become problematic. The use of current values to calculate the tax liability on vessel, gear and quota can impose a burden of indebtedness on the successor which makes fishing unprofitable. Only those enterprises that have legally restructured themselves through incorporation or a holding company find succession and transfer easy to achieve. In an industry traditionally based on family relations, some firms are unwilling to make the necessary legal changes because of the implied socio-cultural shifts and the potential loss of fishing rights for those members of the household – wives and daughters – not normally engaged in fishing. Because women often have rights in the firm, problems can arise in the event of death or divorce. A widow or divorced wife can seek a settlement based on her portion of the joint property, including the value of the assets tied up in the fishing enterprise. With the added burden of inheritance tax, such changes to the family enterprise can lead to the enforced liquidation of otherwise viable firms. Once again the problem lies in commuting the high value of the fishing rights into disposable capital to meet the tax obligations. In this sense, the trade in fishing rights fails to achieve the requirement of economic efficiency. Business closures, brought about in this way, are also socially inefficient, causing unnecessary destruction of tradition, professional knowledge and employment. Not only are the livelihoods of the seven or eight crewmen sacrificed with each business closure, but there is a knock-on effect in shore based employment. Such losses are to be set against a healthier structure for the industry with smaller number of vessels with adequate fishing rights and better prospects for the long-term sustainability of the industry.

Scaling down of enterprises

For the owner of more than one vessel an adjustment of the individual quota is a clear possibility by scaling down the business from two vessels to one and accumulating the fishing rights on the surviving vessel. This unfortunately involves the loss of employment for the crew of one or other vessel and as these are often family members or neighbours the decision is a difficult one. But it happens, as in the case of the owner of two vessels who came under pressure to dispose of one of the vessels:

> 'Fortunately it worked out for everyone from the old vessel. Now we have one ship with a satisfactory quota and we also have our peace of mind back.'

Scaling down is another consequence of the negotiability of quotas. One fisherman, wanting to change from a large vessel with 1600 HP and an insufficient quota to a smaller vessel with 300 HP, commented

> 'Yes, I want to go smaller. Fit the boat to the quota and then you may always fish. Less overheads and the total costs are reduced: less oil, nets, chains, everything.'

Scaling down can also result from the death of a partner where traditionally the wives of fishermen have the same rights in the enterprise as their husbands. When a widow demands her portion, liquidity problems can ensue, leading to closure or the transfer of the business to a smaller vessel.

Although it may seem logical that scaling down would benefit the fish stocks, this is not necessarily the case. In the theoretical model of a pollution licence market, it is hypothesized that the costs, coupled with the pollution effects, would promote interest in the development of environmentally friendly technology. At one level, the switch from a large to a small vessel is beneficial in terms of lower fuel consumption and smaller catches, as well as being cost saving for the enterprise. But scaling down, on a large scale, will tend to increase fishing effort within the coastal waters of the 12 mile zone:

> 'The twelve miles are set to protect the fish. But regarding the intensity of fishing by the smaller Euro-vessels, there has never been so much fishing in the twelve mile zone as in the last few years. We never previously fished inside the twelve mile zone.'

> (Dutch sole and plaice fisherman)

As the breeding grounds for several stocks lie within the twelve mile zone, this new form of overcapacity has potentially serious effects for the environment.

In summary, four socio-economic effects can be identified:

- the number of large and medium sole and plaice fishing enterprises is decreasing;
- the larger fishing enterprises are gaining in strength;
- at the same time, the number of coastal fishermen is increasing; and
- the traditional structure of the family business is in transition.

The environmental effects are both positive and negative. Positively, the overcapacity of the Dutch fleet is being reduced and the increasing concentration of fishing rights held by a smaller number of large enterprises helps to reduce illegal fishing. But an important negative environmental effect is the growing concentration of small-scale fishing in the coastal zone which can intensify the pressure on a sensitive environment. The environmental problem has shifted and requires new measures to address the situation.

Environment use rights market or government regulation?

Free trade in environmental use rights is perhaps one possibility for lessening the damage to the environment. But is not quota setting itself sufficient to achieve the intended environmental effect? Defining limits to environmental exploitation and allocating quotas to the sum of these limits should, in principle, be enough. But the experience of the sole and plaice fisheries with the earlier 'grey market' and rumours of a black market for illegal catches suggests that the setting of limits and quotas is insufficient. The question, therefore, is whether the opportunity to trade the quotas would provide a sufficient incentive. When a resource user is in possession of adequate use rights, the temptation to fish illegally is removed and the only transgressors are those with insufficient use rights to function efficiently. The opportunity to acquire additional rights through purchase or leasehold should therefore form the basis of a solution.

The consequence of a quota market, with intense competition for the limited stock of environment use rights, is that there are fewer places available within the resource user group. The smaller group can more easily govern themselves; at the same time, having sufficient quotas at their disposal, they are more amenable to management by central Government. But the key question is whether the quota market can take the place of government regulation. As we have seen, damaging effects can occur both to the environment and to the socio-economic structures of the fishing industry, including a redistribution of fishing effort within the inshore waters, loss of employment opportunities, the dissipation of professional knowledge and the destruction of a traditional culture. What is more important? A useful instrument of environmental regulation or an environmental policy with a socio-economic component in which it is not only the strong that can exert their rights? The one does not automatically rule out the other. The market, acting as an instrument of regulation, can be developed as an integral part of an environmental policy framed in terms of economic efficiency and social justice. The market mechanism exerts not only a pragmatic but also a primitive function: a market in environmental use rights must not be seen as an alternative to government regulation but as a supplementary instrument of environmental control.

References

Burg & De Raad (1992) *Fiscale Aspecten met Betrekking tot Quota*. Visserijdag, Rotterdam.

Bressers, J.Th.A. (1985) *Milieu op de Markt. De Controverse Tussen twee Markt-benaderingen in het Miliubelied*. Kobra, Amsterdam.

De Boer & Van Keulen (1992) Memorandum inzake fiscale problemen in de bedrijfstak van visserij. In *Fiscale Aspecten met Betrekking tot Quota* (ed. Burg & De Raad), Visserijdag, Rotterdam.

Hoefnagel, E. (1993) Kettingreacties in het visserijbeleid – via onbedoelde naar beoogde beleidseffecten. *Facta*, **4**, 2–7.

Jorion, P. (1982) All brother crews in the North Atlantic. *The Canadian Review of Sociology and Anthropology*, **19** (4), 513–26.

Kunneke, R.W. (1992) De verdeling van eigendomsrechten als bestuurkundig vraagstuk. *Bestuurskunde*, **4**.

Ministerie van LNV, (1992) Vissen naar evenwicht. *Beleidsvoornemen. Structuurnota Zee – en Kustvisserij*. Ministerie van Landbouw, Natuurbeheer en Visserij.

Nentjes, A. (1988) De economie van het mestoverschot. *Tijdschrift voor Milieukunde*, **3** (5), 159–64.

Peeters, M. (1992) Marktconform Milieurecht? Een Rechtsvergelijkende Studie naar de Verhandelbaarheid van Vervuilingsrechten. Tjeenk Willink Zwolle.

Salz, P. (1991) *De Europese Atlantische Visserij: Structuur, Economische Situatie en Beleeid*. Onderzoekverslag, **85**, LEI-DLO.

Chapter 7
Marine Management in Coastal Japan

ARNE KALLAND

From the early seventeenth century people's perceptions of the sea developed very differently in Western Europe and Japan. Whereas the ideology of the *Mare Liberum* spread in Europe after Hugo Grotius published his influential book in 1609, the Japanese authorities closed the sea to everybody not residing in settlements defined as fishing communities. Each village was given a territory to which only its inhabitants had access. At the same time, a licensing system was introduced in many parts of the country and this limited access to marine resources still further.

It is possible to trace the development of the Japanese marine management system through several centuries and one is struck by the continuity of the institutions. In few other countries have customary regulations been incorporated into formal law as in Japan. The coastal waters of Japan have for centuries been under co-management regimes (Pinkerton, 1989) with fishermen participating in the formulation of fisheries policies, allocating fishing licences and playing active roles in the implementation and enforcement of regulations (Short, 1989). Japan's experience is therefore of particular interest to social scientists, fisheries administrators and others who study marine adaptation and management systems.

The Japanese system of sea tenure, which briefly can be defined as 'the ways in which fishermen perceive, define, delimit, "own" and defend their rights to inshore fishing grounds' (Akimichi & Ruddle, 1984; Cordell, 1984), is based on the two principles of fishing rights and licences. Fishing rights define access to a particular space of water while licences give the holders permission to conduct certain types of fisheries. Whereas fishing rights are limited to the inshore coastal waters, licences are issued both for coastal, offshore and pelagic fisheries[1]. Both rights and licences have developed over the centuries and it will be illuminating to take a brief look at how these institutions evolved.

The development of fishing rights

Although the foundations for the present territorial arrangements were created earlier, it was during the Tokugawa period (1603–1868) that fishing territories were firmly defined. Coastal waters were regarded as an extension of the land and thus an integral part of the feudal domain. The sea could therefore be disposed of as the feudal lords (*daimyō*) saw fit. Although some differences existed between fiefs, most of the lords partitioned and allocated the sea space to fishing communities in return for payment of taxes and *corvée* labour. Hence, the fishing territory emerged as a village estate and came under the control of the village headman (*shōya*) assisted by a few village elders. The boundaries of such territories did not necessarily coincide with the seaward pro- jection of the terrestrial boundaries of the fishing villages, as claimed by Ruddle (1985), but were defined by easily recognizable topographic features such as capes, rocks, estuaries and shallows. Occasionally, particularly inside bays, two or more villages shared a joint (*iriai*) territory. Frequently villages also gained access (*nyūgyo*, i.e. 'entry fishing') to their neighbours' territories.

The years between the collapse of the feudal regime in 1868 and the enactment of the first national fishery law in 1901 was a period of transition whereby elements of the old sea tenure system were salvaged or adjusted to meet the demands of industrialized Japan. With the dissolution of the fiefs in the 1870s the old fishing rights, that had existed during the Tokugawa period, reverted to the central Government and a short period with open access ensued. However, as conflicts between old and new fishing villages escalated and large operators improved their position *vis-à-vis* small-scale fish- ermen, the old system of exclusive village territories under the local authorities (now the prefecture rather than the domain) was re-established. But in 1887 the central Gov- ernment encouraged the establishment of regional fisheries' co-operatives to co- ordinate the use of fishing territories (Ruddle, 1987), and this enabled the co-operatives in collaboration with the prefectural authorities to encode existing prescriptive rights and this codification served as a model when the *Fishery Law of 1901* defined four types of rights [see Fig. 7.1 (a), (b) and (c)]:

(1)   Exclusive fishing rights (*jisaki senyō gyogyōken*) gave the fishermen of a village exclusive access to a territory, which sometimes was shared with a few other vil- lages, whereas joint exclusive fishing rights (*kyōyū senyō gyogyōken*) implied a codification of the old access (*nyūgyo*) rights to territories of other villages.
(2)   Set net fishery rights (*teichi gyogyōken*) gave the holders rights to certain areas for stationary equipment such as set nets.
(3)   Sectional fishery rights (*kukaku gyogyōken*) gave the holders lots to cultivate fish, shellfish and seaweeds.
(4)   Special fishery rights (*tokubetsu gyogyōken*) allowed the holders to operate large nets in coastal waters for species like sardine, mackerel and sea bream, as well as to construct artificial shallows in order to attract fish.

As these rights, to a large extent, were based on practices going back to the feudal past, the most important difference from the Tokugawa system was the transfer of the exclusive rights from villages to local fishing co-operatives which were established throughout the country just after the turn of the century. In fact, both types of exclusive fishing rights became estates of co-operatives. Although this improved the fishermen's influence on management questions considerably – because merchants in their capacity as village leaders, in theory at least, lost their access to these estates – merchants nevertheless retained a strong grip on the fisheries. Many of them played important roles in fish curing and marketing, as net owners and even as leaders of co-operatives. Moreover, individuals could acquire fishery rights in the last three categories, and these tended to become permanent as they were easily renewed and could be inherited, leased and even sold (Ninohei, 1981).

The new fishery law of 1949 sought to remedy this situation. Only active fishermen can become members of the co-operative and, although certain rights can still be allocated to individuals and private enterprises, sectional fishery rights and set net fishery rights can no longer be transferred to other persons. Moreover, individuals and private enterprises will only obtain such rights as long as the cooperative, or a group of fishermen, do not want to make use of their privileges. Such rights are usually given for five years and can neither be leased nor sold without authorization. The fishing territories were also redefined and given over to reorganized fishing co-operative associations (FCAs; *Gyōgyo kyōdō kumiai*). The exclusive fishery rights were transferred into common fishery rights (*kyōdō gyogyōken*) whereas special fishery rights were replaced by licences. Three categories of rights thus emerged (Akimichi & Ruddle, 1984; Ruddle, 1985):

(1)   Community fishery rights are exclusively given to FCAs. These allow the members to operate within the exclusive territory of the FCA provided they keep the regulations stipulated in the rights and hold the necessary licences. There is still a distinction between exclusive rights (*tanyū kyōdō gyogyōken*) and shared rights (*kyōyū kyōdō gyogyōken*).
(2)   Sectional (or demarcated) fishery rights (*kukaku gyogyōken*) are usually given to FCAs, in which case it is up to the FCA to distribute them to its members, but they can also be given to individuals and private organizations.
(3)   The large set net fishery rights (*teichi gyogyōken*) can also be given to individuals, private enterprises and FCAs.

If we take an historical approach to the development of the exclusive community fishing rights, we will discover that there has been a general trend towards larger territories. One of the main problems which has faced Japanese fishermen for centuries is the fact that small exclusive fishing territories based on a village were designed to solve political problems rather than as a means of resource management. The delimitation of the village territory did therefore not take local conditions into account to any significant degree, and the fixed and exclusive fishing territories were in many cases maladaptive from the fishermen's point of view. In the first place, fish are capricious. Except perhaps for some benthic species, one seldom knows where the fish will appear. Migratory species, which in Japan are very important, may change their migration routes. Secondly, many territories cannot support year-round fisheries. Thirdly, small territories make it difficult for the fishermen to specialize within certain fisheries. Though this may help conservation it must be regarded as poor management when fishermen are idle in one place while their neighbours have too much to do. Finally, fixed territories made it difficult for the fishermen to respond to ecological changes as well as to changes in technology, marketing structures, human demography and so on.

For these reasons, many fishing villages have long tried to obtain access to the territories of their neighbours. Some have been inclined to pay fees for this access, while others have tried to obtain the same by strategic poaching or by trading access to each other's territories. In Fukuoka most of the documents about fishing disputes during the Tokugawa period are about gaining access to a neighbour's territorial waters. With time the system became extremely complicated and during the Meiji era (1868–1912) it was encoded in the joint fisheries rights shared by several villages where new concessions for 'guest' fishing were also given (see Fig. 7.2). With the introduction of engine powered fishing vessels it became even more important to enlarge the operational field and, after World War II, the joint exclusive fisheries rights were abandoned and the area opened to

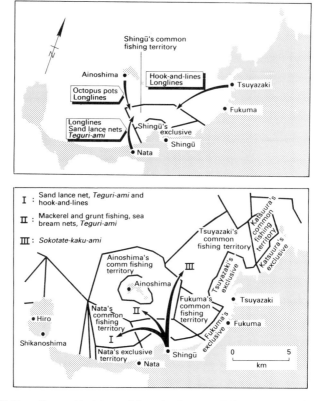

Fig. 7.1 (a) Fishing villages with rights to fish in the Common Fishing Territory of Shingū, 1891. (b) Fishing territories in which Shingū fishermen hold rights to fish, 1891.

all the fishermen from the prefecture (provided they have licences) or the rights have been incorporated into exclusive territories consisting of the amalgamated territory of several villages (see Fig. 7.3)[2]. Fishing rights (and other regulations) can therefore best be seen as continuous processes caused by social, ecological and technological changes (Akimichi, 1984). Illegal fishing – poaching in other villages' territories, fishing in closed areas, using more powerful engines and finer meshes than permitted – is a very common feature and is a part of these processes. Hence regulations do not in themselves necessarily prevent conflicts and might in fact cause conflicts – but they often provide institutional ways to solve them.

Licences

The coastal fisheries were until 1949 mainly regulated by fishing rights. Instead of licensing operations, the authorities chose to limit the number of fishing locations. When there was more gear than allocated sites, the fishermen took turns, decided by rotation, lottery or on a 'first come first served basis'. In this situation, new nets did not increase the fishing effort nor pose a threat to resources, although they might have caused over-capitalization and reduced profits.

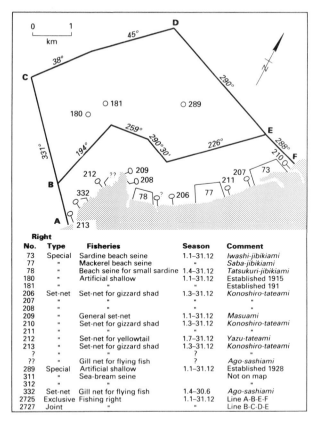

Right				
No.	**Type**	**Fisheries**	**Season**	**Comment**
73	Special	Sardine beach seine	1.1–31.12	*Iwashi-jibikiami*
77	"	Mackerel beach seine	"	*Saba-jibikiami*
78	"	Beach seine for small sardine	1.4–31.12	*Tatsukuri-jibikiami*
180	"	Artificial shallow	1.1–31.12	Established 1915
181	"		"	Established 191
206	Set-net	Set-net for gizzard shad	1.3–31.12	*Konoshiro-tateami*
207	"		"	"
208	"		"	"
209	"	General set-net	1.1–31.12	*Masuami*
210	"	Set-net for gizzard shad	1.3–31.12	*Konoshiro-tateami*
211	"		"	"
212	"	Set-net for yellowtail	1.7–31.12	*Yazu-tateami*
213	"	Set-net for gizzard shad	1.3–31.12	*Konoshiro-tateami*
?	"		?	?
??	"	Gill net for flying fish	?	*Ago-sashiami*
289	Special	Artificial shallow	1.1–31.12	Established 1928
311	"	Sea-bream seine		Not on map
312	"		"	"
332	Set-net	Gill net for flying fish	1.4–30.6	*Ago-sashiami*
2725	Exclusive	Fishing right	1.1–31.12	Line A-B-E-F
2727	Joint	"	"	Line B-C-D-E

Fig. 7.1   (c) Fishing rights in Shingū, 1906–1928.

Nevertheless, licences have ancient roots and have come to play an ever increasing role in Japanese fisheries. Many villages and individuals were given privileges (often monopolies) to use certain gear, as far back as in the Heian period (794–1185), if not before. It was thus an alternative to bestowing fishing rights on loyal followers.

During the Tokugawa period several kinds of licences were in use. For example, in Fukuoka Domain, four (later three) fishing villages were licensed by the feudal lord to dive for the valuable abalone against payment of fees (Kalland, 1987; 1994). Large-scale nets were usually also licensed by the authorities and fishing villages in the vicinity were consulted before a new licence was issued to a village or an individual. The operators of the same gear often formed guilds (*nakama*) in order to protect their interests and exclude newcomers from the fisheries. Although the feudal authorities did not promote monopolies in the fisheries, the guilds often received strong support from the village leadership. The struggles between the licence holders and those outside the guilds could last for years and become very bitter indeed (Kalland, 1984; 1994).

The licences issued during the Tokugawa period were designed to regulate coastal fisheries, usually inside the exclusive village territories. With the development of offshore and pelagic fisheries during the Meiji era, it soon became apparent that regulations were necessary also in these fisheries. The central Government introduced licences for several fisheries after 1909 and since 1945 licences cover almost all kinds of Japanese fisheries (Oka *et al.*, 1962).

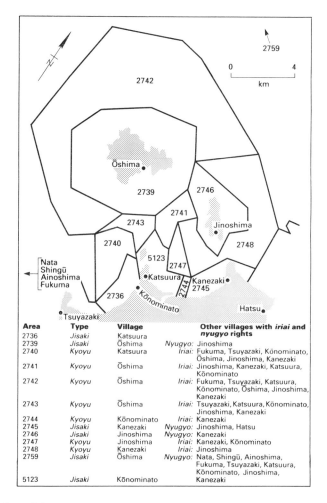

Area	Type	Village	Other villages with *iriai* and *nyugyo* rights
2736	*Jisaki*	Katsuura	
2739	*Jisaki*	Ōshima	*Nyugyo:* Jinoshima
2740	*Kyoyu*	Katsuura	*Iriai:* Fukuma, Tsuyazaki, Kōnominato, Ōshima, Jinoshima, Kanezaki
2741	*Kyoyu*	Ōshima	*Iriai:* Jinoshima, Kanezaki, Katsuura, Kōnominato
2742	*Kyoyu*	Ōshima	*Iriai:* Fukuma, Tsuyazaki, Katsuura, Kōnominato, Ōshima, Jinoshima, Kanezaki
2743	*Kyoyu*	Ōshima	*Iriai:* Tsuyazaki, Katsuura, Kōnominato, Jinoshima, Kanezaki
2744	*Kyoyu*	Kōnominato	*Iriai:* Kanezaki
2745	*Jisaki*	Kanezaki	*Nyugyo:* Jinoshima, Hatsu
2746	*Jisaki*	Jinoshima	*Nyugyo:* Kanezaki
2747	*Kyoyu*	Jinoshima	*Iriai:* Kanezaki, Kōnominato
2748	*Kyoyu*	Kanezaki	*Iriai:* Jinoshima
2759	*Jisaki*	Ōshima	*Nyugyo:* Nata, Shingū, Ainoshima, Fukuma, Tsuyazaki, Katsuura, Kōnominato, Jinoshima, Kanezaki
5123	*Jisaki*	Kōnominato	

Fig. 7.2 Exclusive fishing territories in East-Munakata County, Fukuoka Prefecture, 1901.

Licences for small-scale coastal fisheries are issued by the prefectural government represented by its Fisheries Agency either to a FCA, in which case the FCA allocates the licences to its members, or to individuals and private enterprises. The Fisheries Agency also establishes regulations pertaining to fishing gear, minimum size of the prey, closed seasons and areas, and so on. However, before making decisions on such questions the Agency is required by law to seek advice from the appropriate Sea Area Fishery Regulatory Commission (*Kaiku gyogyō chōsei iinkai*). The jurisdiction of most commissions, which usually have 15 members of whom nine are fishermen elected by members of the FCAs in the prefecture, coincides with the prefectural waters, an exception being Hokkaidō which is divided into ten commissions (Short, 1989). Hence the fishermen have considerable influence on this important body. The central Government issues licences for large-scale pelagic fisheries while the licences for medium-scale offshore operations are issued by the central Government but their number is decided by the local governments. Licences for large and medium scale fishing are mainly issued to individuals and enterprises for a period of five years (Akimichi & Ruddle, 1984).

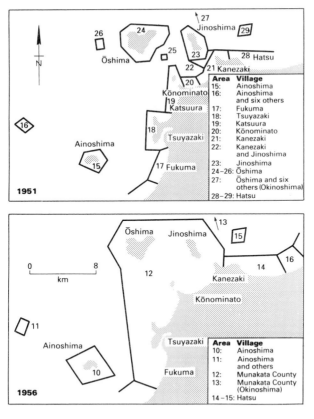

Fig. 7.3   Fishing territories in Munakata County, Fukoaka Prefecture in 1951 and 1956.

One of the important objectives of the management system in modern Japan has been that 'high productivity fishing grounds along the coast should be left entirely to low efficiency boats, while high efficiency boats should be used to open up low productivity offshore fishing grounds which cannot be developed satisfactorily by low efficiency boats' (Oka *et al.*, 1962). The authorities have aimed at securing the livelihood of the small-scale fishermen by placing restrictions on outside investments in coastal fisheries thereby preventing capitalist firms or individuals from usurping the access rights or means of production of the indigenous fishermen. The law has, from this point of view, to a large extent been a success as the great majority of the management units in coastal fisheries are owner operated and small scale, and few licences have been issued to the more capital intensive operators for fishing in the coastal areas. The more efficient vessels have thus been forced to enter the offshore and pelagic fisheries. It is believed that this policy has been a major factor behind the rapid opening of new fishing grounds and thus the increasing total catches. Whereas the coastal fisheries have produced between 2.5 million and 3 million tonnes annually since 1925 (with a slight increase to more than 3 million in recent years due to aquaculture), pelagic and offshore fisheries have increased manyfold since then.

The exploitation of both fishery rights and licences is restricted by a number of regulations. The fishery rights cover only certain species and gear and might also have seasonal limitations. Licences are restricted by regulations pertaining to the size of the vessel, the power of its engine, the number and design of gear, seasons, areas, quotas and so on. Each licence holder receives a certificate which is endorsed if he is found fishing in contravention of these regulations, and the licence can be confiscated after repeated violations.

The Fisheries Co-operative Association (FCA)

Although the responsibility for both fishing rights and licences today ultimately rests with the national Government, represented by the Ministry of Agriculture, Forestry and Fisheries, one of the main features of the Japanese management regime is the delegation of as much of the responsibility as possible to the local FCA. Through the Fisheries Co-operative Association Law of 1948, established in order to improve the economic and social position of the fishermen, and the Fisheries Law of 1949, most of the day-to-day administrative tasks are delegated to the FCA. It is left to its members and board of directors to interpret, implement and enforce national and prefectural fishery regulations (Ruddle, 1987), and it is therefore possible to adjust regulations to local conditions and interests.

The FCA is an indispensable institution in any Japanese fishing community. Not only does the co-operative perform a number of services to its members such as organizing courses and excursions, providing banking services, selling fuel and fishing gear cheaply and marketing catches but it also performs a number of management functions. It holds the community fishing rights as its estate and has first priority in obtaining sectional and large set-net fishery rights as well. Whereas it is the Fisheries Agency that gives fishing licences to an FCA, it is up to the latter to distribute these to its members. It is also the co-operative that allocates particular fishing sites to members and establishes regulations peculiar to that FCA and its fishing territory. Often its internal regulations are stricter than those at the national and prefectural level. Nonetheless, violations of regulations laid down by one's own FCA seem to be rather rare, probably due to the social stigma that such behaviour leads to (Short, 1989). The FCA-controlled territories are thus an important stabilizing factor in the management of coastal fisheries. Finally, the FCA mediates between members of different FCAs and works actively for the enforcement of fishing regulations, particularly of the regulations imposed by the FCA itself.

The law stipulates that only active fishermen – defined as persons who fish more than a specified number of days (usually 90 or 120 days) annually – who live in the FCA's jurisdictional area can become members and hold shares in the FCA. Practice varies but in most areas only the head of the household is a full member (*seikumiai'in*) of the FCA; other male members can become junior members (*junkumiai'in*) without voting rights. The FCA is led by a board of directors (*rijikai*), headed by a president (*kumiaichō*) and a couple of advisers (*kanji*) supplemented by a few regular members (*riji*). Usually all are elected fishermen while the daily administrative work is taken care of by employees responsible to the board of directors. Larger FCAs are divided into sections based on residence or type of fishing gear.

In their FCAs the Japanese fishermen have had a powerful tool by which to influence marine resource management. Moreover, the Fishery Law of 1949 established the fishing rights as a real property right (Wigen, 1989) and the FCA can demand 'fair' compensation for any action including governmental expropriation that infringes upon these rights or reduces their value. This has enabled the fishermen actively to defend their access to marine resources and has placed the FCA in a position where it can block private development projects, such as land reclamation works and construction of factories and harbours. Private developers have to negotiate with the FCA in question to decide the terms of compensation for abandoned fishing rights before any work can commence, and the law requires the written consent of at least two thirds of the FCA's members before such rights can be relinquished.

Negotiations can take years (Befu, 1980; McKean, 1981) and, not surprisingly, the strong position of the fishermen has caused compensation to reach high levels. But it has also created conflicts among the fishermen who want to sell and those who do not. There are clear indications that elderly fishermen without heirs in the trade generally prefer short-term profits to long-term resource management and frequently it is the elderly, senior members of the FCA who want to sell their rights while the younger ones, many of whom are only junior members without voting rights, are opposed (Kalland, 1981; Short, 1989). Moreover, the pressure from the industries and the authorities toward the leaders of the FCA has led to cases of corruption (Okada, 1979; Befu, 1980) and FCA leaders have also been accused by rank-and-file members of taking an unreasonably large portion of the compensation (Befu, 1980). The outcome is usually the same however; the FCA gives up some of its rights in return for huge compensation. In the short run, the fishermen benefit financially (and many return in fact to fisheries in order to get part of the compensation money) but these outcomes might be detrimental to the development of coastal fisheries, in the longer term.

In some ways, then, the exclusive fishing territories have generated conflicts and might have been an obstacle to the development of the coastal fisheries. It is, however, premature to conclude that the exclusive territory today is an anachronism as some have done. There are, in particular, two fields where the fishing rights may play increasing importance. One is to regulate the cultivation of fish, shellfish and seaweed. The FCAs and the sectional fishing rights are useful tools for securing this industry for the local fishermen, and aquaculture has become an important secondary source of income to the Japanese fishermen. The ownership structure in aquaculture is thus strikingly different from that found in many western countries where larger commercial enterprises tend to dominate.

The other field where the FCAs and the system of fishing rights can be useful is to protect the fishing grounds against environmental deterioration. Although the fishermen have been unable to prevent pollution disasters like the ones in Minamata and the Inland Sea, the FCAs have been able to negotiate large compensations for loss of fishing grounds due to pollution and have also been able to block new polluting projects. It should also be mentioned that many FCAs have done much to improve the coastal fishing grounds, for example by constructing artificial shallows.

The exclusive territories resemble in several ways the new economic zones which give the coastal states exclusive property rights both over the marine life (with certain modification where a stock is shared by several nations) and mineral resources on the seabed and beneath. Few other countries give such fundamental rights to the fishermen.

An assessment

In the West it has been overfishing that has forced the authorities to give up the 'open sea' policy and to limit entry to many of the fisheries, often against the wishes of the fishermen. Fishery experts and law makers have therefore accused the fishermen of being unable to regulate their own fishing. The fishermen should have been victims of 'the tragedy of the commons' (Hardin, 1968), a concept that in recent years has come under severe criticism. Limited entry has, on the other hand, long been taken for granted in Japan but not necessarily in the name of conservation. Although Japanese fishermen had occasionally argued for conservation in the Tokugawa period, the Tokugawa authorities were mainly interested in public order, the physical survival of the fishermen and supplies of food and corvée labour[3].

Order on the fishing grounds was also an important objective of later governments and this was reflected in the Fishery Law of 1901 as well as in the present law. The fishery rights seek to separate gear that easily come into conflict. But gradually the objective of conservation has come to the fore. More efficient fishing technologies such as the introduction of engines and synthetic nets has led to many cases of overfishing. Development of new markets has had the same effect, and often more licences have been issued than was advisable, due to political, economic, social or other considerations. Much of the implementation of the regulations is delegated to the local FCA which may have different priorities from those of the central Government[4]. But there can be no doubt that conservation has become more important in recent years and is now the expressed purpose of many of the restrictions on fishing efforts[5]. This is particularly evident in the management of demersal species monopolized by single FCAs where cooperatives often have imposed stricter regulations than those required by the law. The FCAs have proved less efficient in preventing over-exploitation of highly migratory resources (Short, 1989).

Fishery rights and licences are designed to meet a number of objectives, such as:

- to secure food for the population;
- to secure the livelihood of the fishermen and their families;
- to preserve order on the fishing grounds; and
- conservation.

The Japanese system of sea tenure has been a qualified success, at least when compared to the performance of other leading fishing nations. To a large extent, it has achieved the objective of providing marine food for the nation. This has been possible first of all by the increase of offshore and pelagic catches and a slow increase in aquaculture along the coast[6]. The number of fishermen has, on the other hand, decreased dramatically from more than 1.2 million in the inter-war years to about 400 000 at present and continues to decline, despite the fact that the income of the average fishing household today can compete with those in other occupations. Although the Japanese management system has been unable to prevent conflicts among the fishermen, the FCAs provide an institutional framework for their settlement. No doubt stocks have been depleted, but the sea tenure system has nevertheless contributed importantly to the rather stable coastal catches between 2.5 and 3 million tonnes over the last 60 years[7].

On the surface, the fisheries are more regulated in Japan than in any other country but it would be a mistake to confuse the written law with real life. The regulations are often only vaguely phrased and leave some room for interpretation. Moreover, illegal fishing is usually only lightly punished. The system is thus much more flexible in reality than on the surface and it is precisely this flexibility that makes the system work and be responsive to change. Fisheries rights, licences and other regulations are woven into a complicated fabric. Regulations which have been introduced in the West to solve the problems of overfishing have been used for centuries in Japan, although for other purposes. We have here a unique opportunity to see how such institutions have worked in practice, how they are changing over time and how the fishermen have reacted to them. The Japanese sea tenure system is indeed a challenge to social scientists and fisheries administrators alike.

References

Akimichi, T. (1984) Territorial regulation in the small-scale fisheries of Itoman, Okinawa. In *Maritime Institutions in the Western Pacific*, (Eds K. Ruddle & T. Akimichi), pp. 89–120. Senri Ethnological Studies No. 17, National Museum of Ethnology, Osaka.

Akimichi, T. & Ruddle, K. (1984) The historical development of territorial rights and fishery regulations in Okinawa inshore waters. In *Maritime Institutions in the Western Pacific*, (eds K. Ruddle & T. Akimichi), pp. 37–88. Senri Ethnological Studies No. 17, National Museum of Ethnology, Osaka.

Befu, H. (1980) Political ecology of fishing in Japan: techno-environmental impact of industrialization in the inland sea. *Research in Economic Anthropology*, **3**, 323–47.

Cordell, J. (1984) Defending customary inshore sea rights. In *Maritime Institutions in the Western Pacific*, (eds K. Ruddle & T. Akimichi), pp. 301–26. Senri Ethnological Studies No. 17, National Museum of Ethnology, Osaka.

Hardin, G. (1968) The tragedy of the Commons. *Science*, **162**, 1243–8.

Howell, D.L. (1995) *Capitalism from Within. Economy, Society, and the State in a Japanese Fishery*. University of California Press, Berkeley.

Kalland, A. (1981) *Shingu: A Study of a Japanese Fishing Community*. Curzon Press, London.

Kalland, A. (1984) Sea tenure in Tokugawa Japan: the case of Fukuoka domain. In *Maritime Institutions in the Western Pacific*, (eds K. Ruddle & T. Akimichi), pp. 11–36. Senri Ethnological Studies No. 17, National Museum of Ethnology, Osaka.

Kalland, A. (1987) In search of the abalone: the history of the Ama of northern Kyushu, Japan. In *Seinan chiiki no shiteki tenkai*. pp. 23–33. Shibunkaku Shuppan, Kyoto.

Kalland, A. (1994) *Fishing Villages in Tokugawa, Japan*. University of Hawaii Press, Curzon Press, London/Honolulu.

McKean, M. A. (1981) *Environmental Protest and Citizen Politics in Japan*. University of California Press, Berkeley.

Ninohei, T. (1981) *Fisheries Exploitation in the Meiji Era*. Heibonsha, Tokyo.

Oka, N., Watanabe, H. & Hasegawa, A. (1962) The economic effects of the regulation of the trawl fisheries of Japan. In *Economic Effects of Fishery Regulations* (ed. R. Hamlisch), pp. 171–208. Fisheries Report **5**, FAO, Rome.

Okada, O. (1979) Japanese fisher people's fight against nuclear power plants. *AMPO*, **11** (1), 38–45.

Pinkerton, E. (1989) *Co-operative Management of Local Fisheries*. University of British Columbia, Vancouver.

Ruddle, K. (1985) The continuity of traditional practices: the case of Japanese coastal fisheries. In *The Traditional Knowledge and Management in Coastal Systems in*

Asia and the Pacific, (Eds K. Ruddle & R.E. Johannes), pp.158–79. UNESCO, Jakarta.

Ruddle, K. (1987) Administration and conflict management in Japanese coastal fisheries. *FAO Fisheries Technical Paper 273*. FAO, Rome.

Short, K.M. (1989) Self-management of fishing rights by Japanese cooperative associations: A Case Study from Hokkaido. In *A Sea of Small Boats*, (ed. J.C. Cordell), pp. 371–87. Cultural Survival Inc., Cambridge, Massachusetts.

Wigen, K. (1989) Shifting control of Japan's coastal waters. In *A Sea of Small Boats*, (ed. J.C. Cordell), pp. 388–410. Cultural Survival Inc., Cambridge, Massachusetts.

Notes

(1) Japanese fisheries can be divided into three categories: (a) coastal fisheries (including aquaculture) which are conducted by small boats – often family operated – of less than 10 t within 12 miles from the shore; (b) offshore fisheries denoting operations between 12 and 200 miles with a voyage lasting up to a week: most of the enterprises are medium or small in scale; (c) pelagic or distant water fisheries are operated by large efficient vessels far from Japan, often within the 200-mile zones of other nations.

(2) The old borders between the exclusive villages territories are, however, still recognized in many regions. This shows the ambivalence fishermen have to this difficult question. On the one hand, they want to enlarge their own field of operations; on the other, they want to protect their home waters. There are also frequent conflicts of interest between units using different gear within the FCA (Kalland, 1981).

(3) The fishing villages had to undertake a corvée in connection with the coast guard and marine transport and some villages had to send sea food to the lord's table. Later the need to provide marine products for export to China and whale-oil as insecticide became important considerations for the feudal authorities. The duties levied on the fishing villages and the fishermen were related to the productivity of the fishery rights and the licences issued by the authorities (Kalland, 1994).

(4) For example, Befu (1980) reports that the leaders of the FCAs can exploit the shortage of licences to their own benefit.

(5) Many of the regulations, which were introduced for other reasons than conservation, nevertheless had conservational side effects. For example, by closing the coast in the early Tokugawa period, 90% or more of the population was excluded from exploiting marine resources. The access to marine resources was further limited when the territories became estates for fishing cooperatives with the enactment of the Meiji Fishery Law in 1901. Finally, absentee net owners were excluded with the Fishery Law of 1949. On the top of this, customary prohibitions against women in most fisheries further limited the number of potential fishermen to a small fraction of the country's population.

(6) However, Japan has probably been harder hit by the new 'Law of the Sea' than any other country and has recently been forced to rely more on imports and joint ventures.

(7)   Perhaps the most spectacular collapses are the sardine and herring fisheries. For a perceptive analysis of the socio-political reasons for the depletion of the herring stocks outside Hokkaidō, see Howell (1995).

Part III
Social Institutions and Fisheries Management

Chapter 8
Social Adaptations to a Fluctuating Resource

TORBEN VESTERGAARD

Introduction

Fisheries management has primarily been linked with biological and economic theories and research. But it is obvious that fisheries management has social consequences, and therefore it is also relevant to enquire what those consequences are likely to be and whether social objectives could and should be included in fisheries management, and if so, in what way. The purpose of this paper is to address some of these issues in the context of a naturally fluctuating resource base and from research findings relating to Danish fisheries.

What are the social issues?

Some issues are quantifiable and can be informed by the collection of general data sets, and some issues can be included in formal equations. Relevant issues of that kind could be income levels and distribution, changes in employment, family composition, recruitment to and demographic composition of fishing communities, changing distributions of ownership, etc. Such issues can be dealt with in the continuation of the received research traditions of bio- and socio-economics. Some of the social sciences can readily contribute to this type of generalizing research and, perhaps, to formal modelling informed by collection of standard data sets.

Other kinds of social issues relate to meaning and values in the social life of the fishing industry. Their illumination requires different research methods and depends on other bodies of theory. One expects social anthropology, ethnology and, to some extent, sociology to be among the social sciences occupied with meaning, values, action and social organization.

The relations between meaning, value and social life depend on their particular context. They would need to be reduced beyond recognition to be subject to a quantifying, generalizing analysis; crucial information is invariably lost in the reduction of the meaning of life to measurable indices of job satisfaction. Such qualitative issues are important for the evaluation of compatibility between management systems and the social contexts in which they have to function. The motivations of actors in the fisheries and the moral legitimacy of management measures are closely linked with meaning and values. Social stability, acceptance of policies, and the solidarity of fishermen within the state or EU are preconditions of successful bio-economic management and cannot be taken for granted.

Danish fisheries in the twentieth century

A short example may be taken from recent research on the relations between knowledge, action and social identity among Danish fishermen (Vestergaard, 1990; 1991a, b; 1994). Danish fisheries changed greatly during the twentieth century in terms of both technological and economic progress. Until perhaps ten or fifteen years ago, modernization was generally experienced as change for the better: increasing efficiency, relief from the heavier burdens of manual work and growing prosperity. At the same time the organization of the fisheries did not alter very much: fishing vessels were predominantly owner operated; vertical integration of catch and processing sectors were the exception; crew members remained risk-sharing co-adventurers; prospects for crew to become owners were good; and limitations on open access remained quite rare. The relationship between fishermen's knowledge, their strategies of action and their social identity remained very stable – at least until 10 to 15 years ago.

The fishermen's knowledge of marine resources is not the general type of knowledge relevant to overall stock assessment, but concrete, time- and place-bound knowledge relevant to their own decision making. Such knowledge includes the expectation of stock fluctuations and limited predictability.

Strategies for action have been geared to cope with a fluctuating resource base through switching between different target species, different types of gear and different waters. Income uncertainties in small and moderate sized enterprises have been buffered by supplier credits, the willingness of the family to reduce their levels of spending or to earn supplementary incomes outside fishing and the ability to mobilize cheap or unpaid assistance within the fishing enterprise. Inventiveness and willingness to experiment and to take risks have been seen as highly valued attributes.

The fishermen have been in a position to see themselves as ideal, modern, self-reliant entrepreneurs. It has been a matter of pride to be able to cope independently with difficulties without receiving subsidies. At the same time they have also seen themselves in a more feudal perspective as servants of the national household, living up to expectations by supplying fish and by serving as moral ideals of the heroic, self-reliant citizen. In that perspective, fishermen did not fish for their own sake but for a higher cause in return for honour and recognition.

The introduction of quotas, relying on expected predictability, contradicted the fishermen's accumulated and present experience of the concrete stock situation. Quotas prevent the fishermen from freely using their usual strategies of action to cope with fluctuating stocks and prices. And this also prevents them from confirming their identity as valued examples of self-reliant members of society and as valued servants of the national household. In fact, the strategies of action that used to earn them recognition now criminalize them. Danish fishermen generally find they have morality on their side; they long to be appreciated again as valuable, law abiding citizens in return for acting as responsible entrepreneurs and using their experience, knowledge and judgment in finding appropriate solutions to the shifts in the resource situations.

Implications for management research

These developments raise the question of whether fisheries management could be designed so as to make use of existing adaptations to fluctuating resources, rather than jeopardize its own moral legitimacy by severing the relationships between knowledge, action and social identity prevalent within the fish catching sector.

Further research is needed to illuminate the existing relationships between fishermen's knowledge, strategies of action and social identity (see for example McCay & Acheson, 1987; Cordell, 1989), and to clarify the merits of existing strategies of action in relation to the sustainability and efficient use of fish stocks whose fluctuations can only in part be attributed to fishing effort. In terms of policy relevance, such research would help in the design of management systems which not only command the support and the acceptance of the fishermen but also make possible the incorporation of the flexibility and adaptability present in customary fishing practices.

In the Danish case, on-going research[1] would seem to suggest that management by distributional intervention (quotas, licences) generates more resistance than management by technical and other non-social means. This relates to the circumstance that distributional intervention interferes with key features of the social organization, whilst technical intervention does not (or does so only indirectly). Technical regulations do not discriminate between persons, but between actions and means. Thus, while everybody must observe restrictions nobody is excluded from participating. One important precondition for the widely held acceptance of a competitive order is that the Danish fishing community does not suffer internally from serious social cleavages. In line with the fishermen's wish to be left in peace while playing the game, there seems to be a widespread antagonism towards management systems which involve external interference with the fishermen's planning and decision making during the fishing year.

Common management of different fisheries

Research on the socio-cultural dimensions of an industry tends to bring local, regional and national differences into the foreground. This is not just because socio-cultural approaches tend to contextualize rather than generalize, but rather because there are important social and cultural differences in addition to the more familiar differences of scale and forms of enterprise. The implications of this would be that a common management policy must either be very complicated in order to take account of the differences within and between national fisheries, or it must be very simple so as to allow room for such differences within a common overall policy.

Historical processes may generate complicated systems that do, in fact, work. For bureaucratically designed systems, however, simplicity would seem to be the safer option. This could be interpreted as an argument in favour of the introduction of individual transferable quotas (ITQs) because they contain some of the qualities of simple systems in their reliance on market forces as the regulating mechanism. The question is whether we can really do better than that. ITQ systems have serious drawbacks and unintended consequences; under such systems differential success in competition becomes accumulative and a threat to continued competition. ITQs also involve the expropriation of commoners from the commons. Yet ITQs still have to rely on the artificial fixing of quotas.

The regulatory effects of existing practices

Research on the relations between knowledge, action and meaning in a particular fishing community often modifies the simplistic image of automatic resource depletion by fishermen, in the absence of private property rights or public control. Strategies of action, following from social constraints and local knowledge, often turn out to have intended (or unintended) regulatory effects on access and effort. An important research topic would be to find out if, and under what circumstances, the modes of practice among the resource users may have regulatory effects which could be usefully incorporated into the management system.

A number of biologists have recently claimed that fish stocks behave chaotically, that is in a deterministic, but unpredictable way (see for example, Smith, 1990; Wilson *et al.*, 1991; Wilson & Kleban, 1992). If this turns out to be the case, it will have far reaching consequences for management systems based on scientific prediction. Furthermore, it will make it increasingly important to study the long-term experiences of fishing communities in coping with fluctuations, how this experience has been accumulated in knowledge and practice, and the extent to which it has adaptive advantages that management systems could take into account.

Anthropologists are sometimes accused of being political or partisan in their research; and they are often seen to side with the people they study. Nevertheless, one should take care not to confuse the selection of key issues with partisanship. Anthropologists obviously focus more on social and cultural viability than biologists and economists, who focus on biological and economic viability. It is a serious error to believe that social and cultural priorities are luxuries in comparison with biological or economic priorities. The moral legitimacy, the social order and political stability of any regime, even a management regime, depend on social and cultural aspects of life.

References

Cordell, J. (ed.) (1989) *Cultural Survival Inc.*, Cambridge, Massachusetts.

McCay, B. & Acheson, J.M. (eds) (1987) *The Question of the Commons. The Culture and Ecology of Communal Resources.* University of Arizona Press, Tucson.

Smith, M.E. (1990) Chaos in fisheries management. *Maritime Anthropological Studies*, **3** (2), 1–13.

Vestergaard, T.A. (1990) The fishermen and the nation: the identity of a Danish occupational group. *Maritime Anthropological Studies*, **3** (2), 14–34.

Vestergaard, T.A. (1991a) Are artisanal fisheries backward? Artisanal fisheries in modern society, the example of Denmark. In *La Recherche Face à la Pêche Artisanale. Symp. Int.* (ed. J.R. Durand, J. Lemoalle & J. Weber), tome II, pp. 781–8. ORSTOM-IFREMER, Montpellier, France.

Vestergaard, T.A. (1991b) Living with pound nets: diffusion, invention and implications of a technology. *Folk Journal of the Danish Ethnographic Society*, **33**, 149–67.

Vestergaard, T.A. (1994) Catch regulation and Danish fisheries culture. *North Atlantic Studies*, **3** (2), 25–31.

Wilson, J.A., French, J., Kleban, P., McCay, S., Ray, N, & Townsend, R. (1991) Economic and biological benefits of interspecies switching in a simulated chaotic fishery. In: *La Recherche Face à la Pêche Artisanale. Symp. Int.* (ed. J.R. Durand,

J. Lemoalle & J. Weber), tome II, pp. 789–802. ORSTOM-IFREMER, Montpellier, France.

Wilson, J.A. & Kleban, P. (1992) Practical implications of chaos in fisheries: ecologically adapted management. *Maritime Anthropological Studies,* **5** (1), 67–75.

Note

(1)   Research as part of an inter-Scandinavian project 'Common Property and Environmental Policy in Comparative Perspective' supported by the 'Nordic Environmental Research Programme 1993–1997', NOS-S and the Danish Research Council for the Humanities.

Chapter 9
Ancient Institutions Confronting Change: the Catalan Fishermen's *Cofradías*

JUAN-LUIS ALEGRET

Introduction – Spain's entry into the European Union

Spain has been a member of the European Union since 1987. In the long process of negotiations which preceded its incorporation as a full member, one of the most difficult topics to negotiate was that of the sea fisheries. The reasons for these difficulties were many and varied:

- the vastness of the Spanish fishing fleet in comparison to those of the other member countries;
- the historical fishing rights of the Spanish fleet in certain areas now pertaining to the Union;
- the importance of the Spanish fish market and a food culture which makes Spain one of the principal consumers and importers of fish in the European Union.

All these factors made for a long and complex process of negotiation, culminating in an Adhesion Treaty that has led to a paradoxical situation concerning the fishing sector; eight years after formal incorporation into the Union, Spain still does not have equal rights with the other countries which form the common fishing market, the so-called Blue Europe.

The majority of agreements reached in this transitional process make reference to the new conditions of access for the Spanish fleet to the Union's fishing grounds. They also refer to other subjects such as:

- the reduction in size and restructuring of the Spanish fishing fleet;
- reduction in fishing effort;
- reorganization of markets;
- free circulation of fish products, etc.

A great deal of information on their importance and, above all, on the problems of applying these norms within the EU, has been published in the press mainly concerning the conflicts between different countries and different fleets and over the use of different kinds of gear. However, a considerable part of these agreements could not be applied *in toto* to Spanish maritime space or to the Spanish fleet. Indeed, areas such as the Mediterranean saw themselves in part excluded when these agreements were put into practice, since the Common Fisheries Policy (CFP) of the EU had still not been implemented, and the member states had not, at that time, extended their exclusive economic zones out as far as 200 miles.

From the moment of joining the EU, Spain became part of Blue Europe, with all the consequences that this entailed. However, with regard to the regulation of access to resources, and the rules covering fishing gear (and given that no CFP existed for the Mediterranean Sea), Spain continued to apply its own norms along with the other Mediterranean coastal members states. For this reason the fleets which operate in the Mediterranean have so far remained unaffected by the restriction of access to their traditional fishing grounds outside the 12-mile limit. Nor have limits been imposed upon the size of the catch, as quotas, or other similar output restrictions, do not yet apply.

New rules of commercialization

The factor which has most directly altered the activity in the Spanish Mediterranean fishing sector, and more specifically in Catalonia, has been the free trade in fishery products and the correlated restructuring of markets and prices. The immediate consequence of this has been the appearance of new rules in the process of commercialization. These new rules greatly affected the whole of the sector, penetrating well beyond the market itself. According to all forecasts, one of the aspects in the Catalan fishing sector which should have suffered a major transformation, apart from the market, was the traditional fishermen's organization: the *cofradía*. The reason is that these organizations do not formally conform to the model of producers' organizations (POs) introduced by the EU, and which serve as a pattern of association through which to structure the market.

According to European Union legislation a PO is 'any organization constituted by initiative of the producers with the aim of adopting the measures necessary to guarantee rational exercising of fishing activities and the improvement of the conditions of sale of its production' (EU Regulation 3796/81). According to these same norms, for a PO to be recognized, certain requirements must be complied with. For instance, it must provide evidence of sufficient economic activity. It must also exclude, within the limits of its jurisdiction, all discrimination against nationality or place of establishment; it must possess the necessary juridical capacity, etc. (Article 2 Regulation of 3796/91).

The *cofradías* are, however, ineligible for subsidies to maintain the minimum withdrawal price of certain pelagic species such as the sardine since, according to the regulations, the *cofradías* do not comply with the minimum requirements for guaranteeing a correct usage of the funds.

Institutional reform

As a result of this new situation in relation to the *cofradías*, created by the incorporation of Spain into Blue Europe, a campaign was carried out within part of the Spanish fishing sector with the objective of making the *cofradías* disappear, or, at least, of transforming them into POs. In this way the *cofradías* could adapt to the conditions laid down by the new Union order. However, if an analysis is made of the different kinds of arguments which were used to justify this objective, it can be seen that they went beyond the strictly juridical aspects of the application of Union norms, and entered the territory of political, economic and social considerations, revealing that interests other than those being professed were in fact the real objectives of this campaign.

The arguments which were given in order to justify this necessity for institutional reform were many and varied. It was claimed by different parties, and in different ways, that:

- the *cofradías* were public law institutions and therefore not independent from the State;
- they were co-operative organizations and therefore not democratic;
- they were organizations with a structure which cannot adjust to the efficient model of POs proposed by the EU;
- they were organizations which do not represent the interests of the merchants; and
- they were a relic from the past that must be overthrown in the name of modernity.

But these arguments, besides revealing a clear political purpose, in many cases only demonstrated considerable ignorance of the true social dimension of these organizations. In this sense, it is not surprising therefore that the campaign, which would have resulted in the disappearance of the *cofradías*, met with complete failure, despite the political pressure which was placed upon them. The proof of this is that not one single PO is functioning in the Mediterranean sea-fishing zones at the present time, despite the fact that this is the organizational model proposed by the Union, and the only one which was defended by the Spanish Government. The setting up of POs for the offshore fisheries had been proposed by the Ministerio de Agricultura, Pesca y Alimentacion (MAPA) in 1986 as part of a strategy for adaptation to future EU norms. In principle 40 POs were to be created for the whole of Spain; of these six were to be located on the Mediterranean coast.

The *cofradía*: an enduring institution

At the present time, the only fishermen's organizations on the Catalan coast with the capability to represent the interests of the fishing sector as a whole are the *cofradías*. Several other organizations do exist including the Vessel Owners Association, trade unions and co-operatives. But these associations do not carry enough socio-political weight to represent the interests of the whole fishing sector by themselves, and thus the *cofradías* are the only organizations with recognized representation and legitimacy (see Alegret, forthcoming). The first conclusion that we can draw from this situation is that any intended institutional restructuring of the Mediterranean fishing sector both generally in Spain and, more concretely, in Catalonia should include, at least, the participation of these established organizations.

Formally the *cofradías* were Public Law institutions, which pre-supposes that they depended directly upon the State for their creation, internal structuring, and the dissolution and election of their management boards. They are regulated by the State through the RD 670/1978 and in Catalonia they now depend directly on the Autonomous Regional Government (Generalitat de Catalunya) in accordance with the Central Government RD 1137/1978 and the Regional Government D152/1992. From these characteristics it may be inferred that the *cofradías* have the same kind of structure, and function in the same way, as ancient guilds or corporations. However, the legitimacy of their representative function is obtained from, and maintained by, diverse sources other than the formal legitimacy granted to them by law.

The *cofradías'* principal source of legitimacy, and therefore the strength of their representative nature, comes from the fact that, formally, they are co-operative organizations recognized by the law (Alegret, 1990). The main characteristic is that they represent simultaneously both parts of the production sector: the social part (the workers) and the economic part (the shipowners). However, this peculiarity becomes one of the first obstacles in the process of compulsory transformation of these entities into POs.

On a second level, the need for restructuring the associative network in the Catalan fishing sector is justified fundamentally from two points of view.

The first argument makes reference to political aspects, more specifically to the need to eliminate State intervention in the associative fishing sector. Organically, the *cofradías* depend upon the State and therefore have certain rights and obligations towards the State, since there are 'organs of consultation and collaboration with the Administration on subjects of general interest referring to extractive fishing activities and commercialization' (RD 670/1978). From this perspective, it is argued that the direct intervention of the State through the *cofradías* is incompatible with the regulation laid down by the EU.

The second argument makes reference to the need to conform to the demands of the EU on subjects such as the democratization of organizational structures, liberalization of the market, and free access to resources for all members of the EU, as well as limitations on extractive activity imposed by the CFP.

Cofradías and producers' organizations compared

In order to reach an understanding of the reasons for this failure in the implementation of POs in the Catalan fishing sector, it is necessary to identify some of the fundamental differences which exist between the *cofradías* and the POs. Through this comparison we will be able to see how any movement towards institutional re-structuring which is attempted without taking into account the social characteristics of these institutions, will not only be doomed to failure but will cause more problems than it attempts to solve.

Between the *cofradías* and producers' organizations there exist certain basic structural and functional differences which may be summarized as follows.

First, POs are organizations which exist due to the initiative of the producers while the *cofradías*, as Public Law Corporations, are created in law by the State. This supposes that the *cofradías*, being subject to public law, cannot be created or run as private organizations. Because of this, the European Union's requirement for the constitution of organizations through the initiative of the producers is not complied with.

Second, *cofradías* are organizations composed of workers and vessel owners, while POs are organizations constituted by producers. According to EU legislation, fish producers are those persons or organizations which market the products which derive from fishing. This is because the POs are regulated within the framework of the organization of the Common Market, and for this reason, when we speak of producers we refer to economic agents who intervene in the commercialization of fishery products, and not to the social actors related directly with the fishing activity.

Third, POs are organizations with juridical capacity for adopting certain measures which will perforce be complied with by their members. The *cofradías*, on the other hand are merely consultative organizations, since it is the State which decrees the aforesaid measures, and these are normally universal in character.

Fourth, *cofradías* are organized according to geographical areas, corresponding to the territorial limits of residence and influence of the fishing communities. The POs, on the other hand, are organized within 'material boundaries'; in other words it is the aims of the association which impose limits and limitations, such as the amount of certain species of fish to be caught, according to the targets of the association.

Finally, the *cofradías* are organizations with a functional structure intended to defend the social interests of the whole fishing community (regulation of access to resources, resolution of conflicts, organization of first sales for all vessels, organization of common services, etc.), which is why they are organized according to geographical areas. The POs, by contrast, have a structure intended to defend only the economic interests of a specific group of producers and dealers, which is why they are not organized according to territorial areas and why they cannot exclude any potential member for territorial reasons alone.

The ideology of co-participation

It comes as no surprise, therefore, that any attempt at institutional restructuring which would transform the *cofradías* into POs will provoke the disappearance of the principal element of socio-political integration which exists in the Catalan fishing communities today, and which reveals itself through an 'ideology of co-participation' in the public affairs of fishing activities, shared by all the members of the community.

This ideology is embodied in, amongst other things, the very existence and defence of the *cofradías* by the whole of the community, as well as in the individual loyalty to the organization expressed by each and every member of the community. This ideology of co-participation may be seen in some of the current characteristics of fishing activities in Catalonia and, more specifically, in the socio-political and economic roles that these organizations play.

(1) The *cofradías* are *de facto* and *de jure* representatives of all the fishing fleet. This fact means that the *cofradías* cannot represent one particular segment of the fishing fleet and must act in the representation of the interests of all segments, simultaneously. The immediate consequence of this is that between the different segments a feeling of belonging to a single group of shared interests is generated, independent of the segment to which the individual belongs. Membership of this group is founded upon the opposition between fishing and non-fishing interests, whether the latter be construed as the central or regional administration or whatsoever other group not directly dedicated to fishing and which, therefore, is not a member of the fishing community nor of the 'brotherhood' (a literal translation of *cofradía*).

This feeling of belonging to one group of common interests is what causes the *cofradía*, and more specifically, its Patron Mayor (President), to be considered as the ideal mediator for the solution of conflicts or problems which arise between the different parts of the group. The viability of this function is demonstrated by the low incidence of recourse to the judicial system for the solution of conflicts.

(2) On the boards of directors of the *cofradías* there is equal participation from both the work force and the vessel owners. This fact, which is determined by law, generates among crew members and owners a feeling of belonging to the same organization, recognizing that the *cofradías* represent jointly the interests of both 'constituencies' and

is therefore the most suitable organization to mediate in the resolution of conflicts which may arise between them.

(3) *Cofradías* exercise direct control over the first-hand sales of fish through the organization of the fish auction. This produces a feeling of security between crew members and vessel owners in face of the laws of the market; this mutual security is translated into the control that the *cofradías* exercise over the auctioning of the catch. At present, the auction markets in Catalonia – as elsewhere in Mediterranean Spain – are under the direct control of the *cofradías* as assigned to them by the State. This is in point of fact an administrative concession that could be assigned to other public or private institutions. Control of first hand sales by the cofradías is considered by the fishermen to be the best system for defending their interests against those of the merchants. In addition *cofradías* are the principal means of financial assistance, which further reinforces the idea of belonging to a single community through membership of *cofradía*.

(4) The territorial identity of the *cofradía* further underpins the sense of belonging to a single professional community, all parts of which conduct their roles within the production process within the defined territorial ambit of the *cofradía*, independent of whether the person is a vessel owner or crew man, or belongs to one sector of the fleet or another.

(5) Within the fishing sector, the prevalence of the share system of remuneration establishes a very different kind of working relationship to that found in any other sector of the economy. This generates a feeling of co-participation between crew members and vessel owners in the profits accruing from the production process, which allows them to justify both their low mean annual real income, as well as its insecurity and the haphazard nature. Finally, it also goes some way towards explaining the inability of trade unions to penetrate an area in which the logic of contractual salary relations is not the principle around which labour relations are organized.

Conclusion

One conclusion that may be drawn from the above discussion is that no attempt to re-structure aspects of the current organizational system of the fishing sector in the Catalan region of the Mediterranean can be undertaken without affecting the system as a whole. This might even lead to the total de-structuring of the sector and the surfacing of a range of conflicts which at present are contained and controlled within the existing organizations.

The principal manifestation of this hypothetical de-structuring could arise from the fact that, if the *cofradías* as such ceased to exist, the sector might cease its policy of self-regulation in all those aspects in which it is currently active. This self-regulation has great socio-political importance, since it concerns such areas as the organization of access to resources, the establishment of fishing zones and timetables for leaving and entering the fishing ports, the organization of first-hand sales through auctions, the collection of taxes, the authorization for ships to establish their base in a certain port, the elaboration of statistics and, perhaps most important of all, the resolution of conflicts from within the organization, without the need to appeal to the jurisdiction of external agents.

In the light of all this, should a process of transforming the *cofradías* into producer organizations occur, it seems logical to conclude that the social and political roles of the former could not be taken over by the latter; and this would force the State to intervene directly in the regulation of all those aspects which the 'laws of the market' do not cover. If this situation were to occur, it is obvious that the ensuing social, political and ecological costs would be astronomical in comparison to the problems which the process was intended to solve.

Perhaps for this reason the *cofradías* will continue in existence, despite the compulsory nature of the new European 'logic', and the efforts made to transform them into POs. Perhaps for this reason also, the new CFP for the Mediterranean, rather than starting from the premise that all institutions which do not conform to its model must, a priori, be restructured, should take into account other kinds of solutions. This would involve the adaptation of the model to suit the historical, political, and social conditions which the current institutions merely reflect – institutions which manifestly fulfil functions that are possibly more important than strictly economic ones. For this same reason it would seem wise to remind those responsible for the formulation of European Union policies that no viable economic union will be possible except where appropriate social and political conditions are created. Moreover, sea fisheries will not prove the exception to the rule.

References

Alegret, J.L. (1990) Del corporativismo dirigista al pluralismo democrático: las Cofradías de Pescadores de Cataluña. *ERES Serie de Antropologia,* **II** (i), 161–72, Museo Etnografico, Cabildo de Tenerife.

Alegret, J.L. (forthcoming) Co-management and legitimacy in cooperative fishermen's organizations: les Confraries de Pescadors de Catalunya, Spain. In *Proceedings of the World Fisheries Congress* (eds R. Meyer, B. McCay, L. Hushak, E. Buck & R. Muth). Athens, May 1992.

Chapter 10
Leviathan Management or Customary Administration: the Search for New Institutional Arrangements

Serge Collet

The swordfish fishery: a tragedy of appropriation

The emergence of a crisis of management concerning a restricted access marine resource, where there is no evidence that the mode of exploitation was itself directly responsible for the crisis, leads us to question the validity of studies which lay claim to Hardin's paradigm of the inevitable tragedy of the commons (Hardin, 1968).

Yet this is precisely the unexpected outcome in the case of the swordfish fisheries of southern Italy. Although the dynamics of the depletion of the swordfish (*Xiphias gladius*) in the Mediterranean have never been properly assessed until now, the driftnet fisheries have been subject to no fewer than 25 legislative decisions since October 1989. In its decree of 22 May 1991, the Italian Government set new limits to the maximum length of the drift net at 2.5 km. Such highly restrictive rules, that overturned an earlier decree (30 March, 1990) by which the maximum length had been set at five nautical miles, were also adopted in European legislation (Council Regulation 345/92). In three years, the principal index of fishing effort had been reduced in law to one sixth of its original level, thus reducing the efficiency of more than half the Italian fleet of 682 trawlers operating throughout the Mediterranean basin but with bases mainly localized within Calabria and Sicily which together accounted for 84% of the fleet. As a result, an economic sector, which had developed rapidly after 1975 and which for many fishing communities had represented the most favourable source of income and a secure basis for the local economy, had been put at risk over a period of just four years – without affording the local actors an opportunity to put their case other than through direct action.

In a society said to be 'ruled by democracy and the market', the fishermen of the Mezzogiorno do not seem to enjoy the rights to determine their own existence and actions with the same degree of autonomy as the agricultural entrepreneurs of northern Italy. The swordfish fishermen have been excluded from the socio-political decision making process and this has led to a situation where the central state has sought to impose norms of behaviour on the fishing industry, whose legitimacy can be challenged on two fronts: first, because it presupposes the inability of resource users to understand the need for management of the resources of the marine ecosystem over the long term and, second, because of the level of influence afforded to bio-economic expertise in the decision-making process. In fact, the first results of research into the stock dynamics of swordfish and the ecological impact of driftnets on cetaceans were known only after the decisions to limit the fishery had been taken with great firmness and sanctioned by an agreement of EU member states.

Is there no other way for the 3000 swordfish fishermen than the expropriation of their jobs and their way of life, leaving them to be pensioned off or compensated through social welfare payments – an extension of the 'aidism' which in southern Italy never works except under conditions of clientelism? Is this the final chapter of a story that began 3000 years ago in the Straits of Messina (see Fig. 10.1) – which Plato called the Straits of Scylla after the first savage predator of the *galeotes*, that is to say, sharks, swordfish and dolphins (*Odyssey*, XII, 92–98). The long established commitment to a distinctive way of life, which depends on the exploitation of a particular facet of the marine ecosystems, has created an economic base as well as a deep cultural meaning; the mode of exploitation of the swordfish, known locally as '*the* fish' (*u pisci*), has shaped the particular social identity of the fishermen.

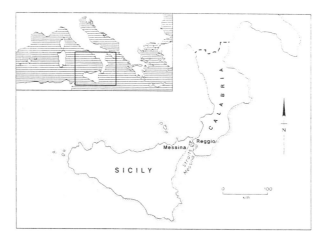

Fig. 10.1   The Straits of Messina.

It should surely prove possible to explore other opportunities for the regulation of the resource, based essentially on local experience and the deep symbolic ties between the fisherman and his 'resource'. Swordfish fishermen have long understood how to regulate their exploitation of nature wisely so as to ensure the social reproduction of the fishermen over the centuries (Collet, 1986; 1993). Why, then, should they not be able today to devise new ways of allocating the resource based on their accumulated wealth of experience and, with the help of social, technical and biological research, develop an appropriate system of management for this common property resource? Why can they not assume responsibility for the control of the resource and the regulation of fishing?

Robustness and flexibility of the marine commons

Night-time swordfishing with the *palamidara* – a 300 m long drift net with 8–12 finger meshes – and hunting swordfish with the harpoon, both practised in the Straits of Messina from boats furnished with very tall observation towers and retractable metal platforms (*passarella*), are the vestigial forms of the oldest maritime culture in the

Mediterranean world. Archaeological research has established that swordfish hunting was practised by the Phoenicians, who frequented the Calabrian area of the Straits of Messina, before the second half of the eighth century BC (Collet, 1995) and were responsible for the expansion of an urban-oriented, sea-based society. Swordfish hunting, directed from a fixed observatory on land where observers stood to watch for the signs of the migrating swordfish at the start of April and shouted warning to the small boats gathered in anticipation of the hunt, is beautifully described in precise detail by the Greek historian Polybius in the second century BC.

The promontory of Scylla marks a strategic meeting point not only for the Mediterranean marine ecosystems, but also between the eastern Aegean and the later north western civilizations. In the course of history, this area had been continuously used as a permanent base for the hunting of sharks (mako, white and thresher shark), killer whales, giant rays, swordfish, marlins and tunas. The long established mode of exploitation was grounded in a dynamic equilibrium between rival harvesting techniques, precisely adapted to the ecology and behaviour of the species and the system of access to the resource which the hunter-fishermen had themselves devised. This system of access, implemented at the start of the sixteenth century AD, provided for an equitable distribution of opportunity based on limited access, the rotation of fishing areas, a drawing of lots prior to hunting and fishing campaigns and fishermen's courts to arbitrate in conflicts caused by the great mobility of their marine prey.

This system of management was developed on the Sicilian coast, where the floating observation posts (heavy, decked *feluca* with a 25 m main mast of pine) were also invented to serve both the original swordfish hunting and, from a later date, the night-time driftnet fishery. A set of simple but effective rules were established as follows:

(1) the designation of those with access rights to the hunting grounds – in 1543 one *feluca* and one *luntre* (hunting boat) made up a hunting crew of 10 persons;
(2) the partition of the hunting ground for the two-and-a-half month summer season;
(3) the drawing of lots to determine the order of entry to the separate hunting territories – the original allocation was subsequently subject to a daily rotation determined by the course of the summer migration of the swordfish along the Sicilian coast; and
(4) the provision that, if the number of hunters with rights of access to the hunting territory exceeds the number of fishing areas available, then the 'extra' fishermen must wait out in the open sea until it is their turn to rejoin the system.

It is not necessary, in the present context, to elaborate the more complicated operational rules to regulate the relations between neighbouring user groups, as for example the right to extend hunting into the territory of another user group, in order to appreciate the simple architecture and wisdom of the system. Suffice it here only to stress the equity and flexibility of a system which uses a 'temporary exclusion order' (*arranti*) to ensure that additional fishermen may take part in the hunt without imposing additional pressure on the common property resource.

In 1777, there were 26 hunting units recorded along the Sicilian coast for an area measuring seven nautical miles in length and half a nautical mile in width; in 1936 there were 88 such units each with an allotted territory 14 km in length. The marine territory – and access to it – is constituted as a social relation which governs competition between resource users for a share of an extremely mobile, wild resource. The possibility of temporary exclusion from the hunting territory was the specific condition, within the constraint of an optimal number of hunting sites along the Sicilian and Calabrian coasts, which enabled the reproduction of a global system at local and inter-local levels and eventually at the level of the individual community which derived its means of existence from the common pool of resources.

Temporary exclusion from the hunting rotation creates, at the local level, optimal conditions for all who take part in the annual hunt on a daily rotational basis. This is reciprocated at the wider regional level, where the Sicilian and Calabrian hunters – and, later, the *palimidara* fishermen – avoid being present on the same fishing grounds at the same time. In Calabria the hunting season lasts from mid-April to the end of June, while on the Sicilian coast – where the swordfish reappears on its migration from the south east – the hunting lasts from early July to mid-September. The absence of one or other of the regional user groups likewise ensures optimal conditions for the hunt by avoiding the overloading of the hunting grounds. The co-operation of each participant with all other co-users is not, in fact, a feature of the hunting process itself – this is, in a sense, a private, individual exploitation of the resource – but occurs out of respect for a set of operational rules which is seen to ensure optimal conditions of exploitation at both local and global levels. At the root of this respect for the system is the acknowledgement and acceptance of a natural law: the swordfish is not in fact present on both coasts of the Straits of Messina at the same time; it is the system, therefore, that ensures that each participant has an equal opportunity for a share of the resource. One other feature which helps to explain the enduring nature of the common property resource was the willingness of the State to accept the fishermen's own system of regulation, and to grant it authority, thus leaving the conflicts which inevitably will arise over the exploitation of a highly mobile and, at times, dangerous resource to be resolved by the hunters' own 'swordfish tribunal' composed of the most expert and generally the oldest harpooners.

Four and a half centuries of successful self-government of the commons should be set against the pseudo-scientific paradigm of the inescapable tragedy of the marine commons. Endogenous technical developments, promoted by the hunters themselves, have certainly intensified the fishing effort, but they have not led to the breakdown of the management system and the ingenious institutional arrangements and social focus that lay at its base. In 1970, the system was recreated with the same success by the fishermen of Alanya in Turkey (Berkes, 1986), permitting the inshore fishery to overcome what Berkes called 'the dark age' of disruptive confrontation caused by the quest for short-term economic benefits. This quest for short-term gains is a key feature in neo-classical economic explanations of the inevitable depletion of marine resources under common property regimes, resulting from a confusion of *res communis* and *res nullius*.

The mode of exploitation of a marine resource described above – an embodiment of the symbolic link between man and a resource considered to be *the* fish and not simply a fish and as such the object of elaborate ritual at its killing (Collet, 1989) – is in principle, a reflection of social or cultural choice and the unconscious structuring of identity. Other fishing gears, deployed in the Straits of Messina, are similarly regulated by the drawing of lots and the rotation of fishing areas – as, for example, the *sciabbiche* (beach seine) or *mutulare* (small drift net) used in the mackerel fishery – but, in terms of their economic significance, they in no way compare with the prestigious swordfishery.

That fishery has inherited 3000 years of prestige earned by the aristocrats of the swordfish hunt, the harpooners. The fact that, two centuries after the invention of the *palamidara* in 1775, the same Calabrian fishing society was able to experiment with new fishing gear – the *spadara*, drift nets with large meshes of 40–50 cm, from 15 to

30 m in height and a constantly evolving length up to 20 km – is the result of symbolic, technical capital. The long-established maritime experience has produced what Bourdieu (1972) has termed an 'assemblage' or *habitus* born of logical practice. For this latest development, there were no technical managers, no goal-oriented process but rather an unintentional rationality.

All over the world, the use of nylon, motor power and electronics has increased and become ever more efficient. It seems that the Government's only concerns are to modernize the fleets in order to improve the return upon investment. Some years ago, the future of the fisheries was seen to lie in industrial fishing or medium capacity artisanal boats, but nowhere in units of low capital input. Under these conditions, the fishermen were no longer in a position to decide for themselves. They were compelled to enter the industrial, market-oriented mode of fishing simply to pay off the considerable costs that the increasing technical inputs imply: they tend to fish harder, intensifying the depletion of the resource.

From long experience, the swordfish fishermen know that swordfish – unlike tunas or mackerel – do not move in schools, that a certain length of net improves the rate of capture and that the conditions of capture are highly sensitive to weather conditions. In the 1970s, when the markets for fresh marine products were enlarged, the swordfish fishing activity became dissociated from its original area to mutate into an industrial fishery. The same fishermen became the promoters of the expansion in swordfish fishing in the Mediterranean, exploring new grounds at an ever increasing distance from the Straits of Messina.

But it is *u pisci* (*the* fish) that they hunt, and no other. They teach their art (*mistieri*) on their travels throughout the Mediterranean, confident that the time will never come when the right to hunt the swordfish is denied them. Often located on the margins of the communities where they live, they frequently express pride in their own success, achieved through hard work and sacrifice, in building their boats and setting up homes for their families without recourse to 'outside' help. And looking back to the times of great poverty some 30 years ago, they find contentment in their ability then to support their families and provide an education for their children that they never had. In much the same way as the swordfish had supported their forefathers in the past, they now need to pursue the hunt further afield – if necessary to the 'ends of the earth' – just as, at other times, other Mediterranean civilizations had done.

The genesis and consequences of new regulatory norms

Up until the end of July 1989, the drift-net fishery for swordfish had been subject to very few regulations (1882, 1959, 1979). In 1989, however, international public opinion, informed by international organizations for the protection of marine fauna, had been shocked by the slaughter of cetaceans caused by the 40 km drift nets used in the Pacific tuna fishery. In December 1989, the General Assembly of the United Nations adopted a resolution on driftnets and on 30 June 1992, a moratorium was introduced, excluding those areas under the jurisdiction of an international agency such as the Mediterranean where regulatory measures were to be adopted based on statistically documented scientific studies.

Already in October 1989, three research studies were planned in Italy, designed to assess the selectivity of the fishing gear, the ecological impact on protected species and the state of stocks of swordfish. They were to be implemented in the spring of 1990 in the context of the Third Triennial Fisheries Plan undertaken by the fisheries division of the Department of the Merchant Marine.

Against a background of public outcries, provoked by media campaigns in support of allegations that the fishermen were responsible for the deaths of 7000 dolphins each year, together with several hundred whales, two executive orders were issued. The first decree, reducing the maximum length of the drift nets to five nautical miles, was published in March 1990 following an initial assessment of the number of dolphins killed at around 3000 to 5000 per annum (di Natale, 1991). This decree was reaffirmed four months later, in July 1990, by the Regional Administrative Court in Lazio and again as a result of the appeals procedure instituted by the State Council. Ten billion lire have been allocated for compensation to fishermen as a result of the effective closure of the fishery, though this figure should be set against a conservative estimate of loss of earnings in the order of 60 billion lire.

But, on 22 May 1991, the maximum length of the drift net suffered a further drastic reduction to 2.5 km. The *spadara* had now to be moored at a depth of 6 m – an unprecedented technical requirement – as opposed to the 4 m ruling in 1990. During the first five days of August – coinciding with the peak of holiday departures to Sicily – swordfish fishermen came from all over the south to blockade the Straits of Messina. This collective show of strength, while it led to a reopening of the swordfishery, banned from July 18 by the decision of the Regional Administrative Court (TAR) in Lazio, was declared illegal and fishermen accused of involvement in the blockade were given six month suspended gaol sentences in November 1993.

The position adopted by the TAR on two separate occasions in 1990 and 1991 was quite unambiguous. In whatever form, drift-net fishing is illegal and must be banned. While awaiting the implementation of a European Union directive on drift nets, financial provision has been geared towards the retraining of fishermen (Decrees of July 7 1991 and 8 August 1991) and compensation for the scrapping of their boats, that is to say the development of a decommissioning programme similar to that already in force for other EU member states.

Thus, while the biological and technical research projects are still in progress and cannot yet fulfil the function for which they were intended – namely the establishment of a scientific basis for regulation of the fishery – the decision-making processes have gone ahead, swayed by the need to appease public opinion concerning the conservation of marine fauna. Public opinion, in fact, focuses upon whales and dolphins – all accidental by-catches of the swordfishery – but largely ignores the sharks, possibly influenced by scenes from *Jaws*, the Hollywood film image of the killer shark. Meanwhile, the political authorities strive – in 1990, at least – to find a balance between, on the one hand, the need to protect an economic activity vital to the fisheries-dependent regions and their communities and, on the other, the wish to conserve as far as possible a species protected under international conventions. The juridical authorities, however, appear to reflect a public opinion deeply disturbed by the alleged slaughter of a species considered to be a symbol of docility and intelligence.

One of the three research projects concerning the ecological impact of drift nets, concluded in February 1992 and adopted by the Italian Government in July 1992, makes clear that the *spadara* fleet comprising 682 fishing units in 1990 had been responsible for the death of dolphins (*Stenella coeruleoalba*). The maximum number of dolphins thus killed is estimated at 1363 in 1991 out of a stock believed to vary between

100 000 and 250 000 individuals. The catch mortality, therefore, cannot be held to threaten the reproduction of the species. For the sperm whale, where the stock is uncertainly estimated at many hundred individuals, prompt intervention by the fisher- men can rescue the rarely trapped mammals and release them alive back into the sea. However, recognition for such actions presupposes a level of trust, on the part of the authorities, in fishermen who have already been found guilty of crimes against the marine environment by the very powerful 'court' of public opinion; half a million people had signed petitions up to October 1994.

For the management of fisheries based on an assumption of rationality, the main priority is the estimation of the state of the stocks, whenever this is possible and prac- ticable. In modern fisheries biology, other factors are now held to exert some influence on the dynamic equilibrium of the marine ecosystem, including the variation in water temperature, levels of pollution and the adaptation of fish species to fishing methods. Bio-economic models of fisheries management are built on a reductionist view which assumes that future stock size is a linear function of the present size. On what kind of information do the political (and juridical) authorities base their decisions? It is certainly true that data on interspecies dynamics are both difficult and costly to acquire. Research, conducted in the framework of the Third Triennial Fisheries Plan, has shown that the average length of swordfish, measured from the tip of the lower jaw to the midpoint of the caudal curvature decreased by 5% between 1985 and 1991. This, added to the composition of the catch varying between 20 and 24 kg per fish throughout the western Mediterranean basin, 'would seem to denote a general reduction in the stock of swordfish, with a progressive shift of fishing effort to lower age groups as a result of the scarcity of older individuals' (di Natale, 1992). A no less significant finding was that the efficiency of the drift net, not significant up to 3 km, is at its maximum between 5 and 8 km and decreases progressively thereafter.

These new data, unknown to the authorities at the time of decision making in 1991 and still not made public, open the way to other decisions such as the one that closed the fishery in 1992. To survive, the fishermen are forced to ignore the rules which are made without scientific legitimation. On what scientific or technical evidence was the UN's global norm for drift nets made? The official documentation of the European Com- mission is silent on this point. It is as if the recommendation of the Commission to reduce trawling capacity by 20% was anticipated in Italy by the implementation of the Government's regulations on swordfishing.

With drift nets limited to 2.5 km in length, the swordfish fleet of 682 units finds itself in an artificially created condition of overcapacity and inefficiency. The 1991 decom- missioning scheme takes nothing away from the claim from 3000 fishermen that they have been effectively deprived of their skills and of their rights to an existence only barely earned at the cost of five or six months at sea. Moves to other kinds of fishing are not feasible; alternative jobs do not exist. And the technical conversion to long lining for swordfish would paradoxically increase the pressure on the stock of younger swordfish weighing from 3 to 20 kg. Moreover, it would increase dependence on supplies of bait – bluefish imported from Spain – and create a unfair market possibly controlled by the Palermo Mafia.

A tragedy of the commons is occurring but one where the basic condition, identified by Hardin (1969), is in fact absent – namely, the depletion of the resource. The tragedy exists despite the recommendation of the European legislature in 1990 to protect artisanal fisheries in economically disadvantaged regions because of their contribution to the maintenance of cohesion in the socio-economic fabric of the fishery dependent regions (EC Regulation 3044/90).

In Italy, the third section of the Regional Administrative Court of Lazio in July 1991 declared all forms of fishing with drift nets illegal (Edict No. 642/91), including those in use today and for more than two centuries by the fishing communities of the Straits of Messina – the *palamidara* used for swordfish and the *mutulara* used for mackerel between mid-September and mid-November. If this decision, since superseded by the EU's proposal to end all drift net fishing by 1997, comes into force, it will mean the end of a system of resource exploitation which through its careful management regime has not led to any form of lasting ecological damage. The historic testimony of this claim is the fact that hunters and fishermen have exploited the resource for some 3000 years without any impoverishment of the resource base, until the 1970s.

Further evidence has come to light as the result of experimental research, carried out by the Societa Cooperativa Idrobiologica e Pesca in Rome in collaboration with the research unit IRPEM in Ancome, on the acoustic perception of dolphins and the response capacity of their bio-sonars[1]. The experiments were initially conducted in the Riccione dolphinarium using a pool divided into two separate areas by a drift net. The behaviour of two groups of dolphin – one blind-folded, the other not – were observed using a remote-controlled submarine tele-camera under conditions of strict security and surveillance. During the third of these trials, a small bronze bell was attached to the net buoy; the dolphins, alerted by the sound of the bell, veered away from the net. The significance of this observation is that more than two centuries ago, the *palamidara* in the Straits of Messina were equipped with bell buoys intended to warn the fishermen that a swordfish was caught up in the nets. The bells came from two important mountain sanctuaries and were blessed during votive processions consecrated to the Virgin Saint of Seminara, the Blessed Virgin of Polsi. Must such management systems, legitimated in categories of magic-symbolic order perish like so many others (Cordell, 1989) at the hands of Cartesian logic that would make man the 'masters and owners of Nature'? The question merits consideration by those for whom centralized management of marine resources has become something of a creed.

Cooperation: a basic element of marine resource management

The industrialization of fisheries throughout the world, with or without regulation by licences, total allowable catches, ground closures or individual transferable quotas, has led to a dramatic impoverishment of marine resources as well as the destruction of distinctive modes of resource exploitation. By accident or design, many such modes were successful in the sense that man could live at peace with nature (Weber, 1992).

The failure of modern fisheries management systems requires that social and natural scientists should together re-examine the system of relations that Western societies have built with their marine wealth. By failing to invite the fishermen themselves to develop new forms of regulation and allocation of the resources, any policy exposes itself to the risk of failure. If the fishermen do not have a global picture of the dynamics of the species and of the marine ecosystem, they do at least know the location, number and behaviour of the species on their own fishing grounds; without this local knowledge, the fishery cannot be exploited effectively. Their experience of the marine environment – its risks and unpredictability – have induced ways of being and thinking which create the basis for specific forms of social action. It is not by accident that according to Plato (*Laws*, 823b) they were to be excluded from the city's government and from public affairs in general – their world is too uncertain. The farmer can become mayor and councillor, but not those whose activities force them to leave the land, sometimes for long periods of time. The political life of fishermen is occasional: there is the winter *karigi* of the hunters of Alaskan whales (Spencer, 1959) and also the winter meetings of the southern Italian or Spanish *marinas*.

This explains why, in so many fishing communities, it is the women who ensure the regularity of social life (Collet, 1991). The intermittent social networks of the fishermen can be structured on the basis of kinship or occupation but, in the case of fishing, these may be fragmented by competing interests. Competition for the same resource does not imply an absence of co-operation, for otherwise fishing societies would be doomed by the destructive effects of individual interests searching only for their own maximization.

The social form is simply a historical product that can be altered as in the case of the inshore fisheries at Alanya in Turkey (Berkes, 1986). There is no maritime society unable to invent and implement its own rules for resource exploitation and for the sharing of the resource; in one case it may take the form of marine territorial use rights, in another religious taboos and elsewhere the clan, village or community ownership (Akimichi & Ruddle, 1984). Hardin (1968) forgot that man appropriates nature only by means of a social form which regulates access by operational norms and rules that can be very coercive and socially very costly for the transgressor. Faced with an increasing risk of social extinction and aware of the consequences of short-term interests leading to catastrophic internal contradictions, what individual or society would not feel called upon to prevent such a tragedy? The theory of politics elaborated by Hobbes and also by Spinoza (1965, 1966) – and further developed through institutional analysis (Ostrom, 1991) – links together with a major contribution from anthropology, according to which it is not possible to create society without sharing; without the sharing of resources and a universal prohibition of incest, on which the idea of exchange is based. What society of fishermen in a situation where no one can be certain of tomorrow as a consequence of the degradation of the common resource, would refrain from collective action to create new rules and a new contract whereby everyone desists from causing damage to others, because to defend the rights of others is to defend one's own rights?

Once the benefits of the positive cycle of reciprocity have been tested and tried, the cycle will maintain its own reinforcing dynamism and come to impose itself as a normative restraint on freerider strategies. If the benefit to the individual actor is to secure his existence – as it is for all those who share the same activity – then the sanctions against those who are tempted to break the rules can be very severe. This is especially the case in societies where respect and personal reputation are important guiding principles. Cooperation and regulated modes of use can thus replace the state of disorder without the Leviathan state having to intervene. Co-operation in the implementation of regulatory arrangements is the most cost-effective means for society to manage the allocation of its marine resources. Its viability is assured by the agreement of those who depend upon the resource for their survival.

These first principles allow us to contemplate new forms of marine resource management, as well as draw attention to the social gains that governments might achieve by promoting their development (or re-development). There is no 'half way house' on the road to creating responsibility for resource management – an idea which is central to the Common Fisheries Policy. The best control is undoubtedly self-control. Herein lies the essential condition for implementing plural decisions without sustaining the social costs and opening up divisions in the social order. Without involving the co-operative action of fishermen and their institutions in management decisions, there can be no end to the global depletion of fishery resources. Ninety per cent of the world's fishermen live from small-scale, inshore fisheries which yield more than half the world's fish harvest. The governance of marine resources at the local level is a priority – the question is what are the criteria of good governance? Enduring forms of exploitation provide a significant indication in their robustness and success: they have generated proof of their viability and their superiority. But they are being destroyed by the processes of an excessive liberal industrialization.

References

Akimichi, T. & Ruddle, K. (1984) The Historical Development of Territorial Rights and Fishery Regulations in Okinawa Inshore Waters. In *Maritime Institutions in the Western Pacific*, (eds K. Ruddle & T. Akimichi), pp. 37–88. Senri Ethnological Studies No. 17, National Museum of Ethnology, Osaka.

Berkes, F. (1986) Marine inshore fishery management in Turkey. In *Proceedings of the Conference of Common Property Resource Management*, pp. 63–83. National Research Council, National Academy Press, Washington, DC.

Bourdieu, P. (1972) *Esquisse d'une Théorie de la Pratique*. Droz, Geneva.

Collet, S. (1986) *Poisson et les Hommes*. Rapport pour la Division Sciences Ecologiques, UNESCO, Paris.

Collet, S. (1989) Faire de la parenté, faire du sang: logique et représentation de la chasse à l'espadon. *Etudes Rurales*, **115/116**, 223–50.

Collet, S. (1991) Guerre et pêche: quelle place pour les sociétés de pêcheurs dans le modèle des sociétés de chasseurscueilleurs. *Information sur les Sciences Sociales*, **30** (3), 483–522.

Collet, S. (1993) *Uomini e Pesce: la Caccia al Pesce Spada tra Scilla e Caribdi*. (ed. G. Maimone), Collana Universitatess saggi, Catania.

Collet, S. (1995) Halieutica phoenicia I. Contribution à l'étude de la place des activités halieutiques dans la culture phénicienne: point de vue d'un non archéologue. *Information sur les Sciences Sociales*, **34** (1), 107–73.

Cordell, J. (ed.) (1989) *A Sea of Small Boats*. Cultural Survival Inc., Cambridge, Massachusetts.

Hardin, G. (1968) The Tragedy of the Commons. *Science*, **162**, 1243–48.

Hobbes, T. (1968) *Leviathan* (ed. C. Macpherson), Penguin Books, Harmondsworth, UK.

di Natale, A. (1991) Marine mammals' interaction in Scombridae fishery activities: the Mediterranean Sea case. *Fisheries Report*, **449**, FAO, Rome.

di Natale, A. (1992) *Gli attrezzi pelagici derivanti utilizzi per cattura del pesce spada adulto: valutazione comparata della funzionalita, della capacita di cattura, dell'impatto globale e della economia*. Ministero della Marina mercantile, Rome.

Ostrom, E. (1991) *Governing the Commons: the Evolution of Institutions for Collective Action*. Cambridge University Press, New York.

Spencer, R.F. (1959) *The North Alaskan Eskimo: a Study in Ecology and Society*. Bureau of American Ethnology, Bulletin 171, Washinton, DC.

Spinoza, B. (1965) Traité theologico-politique. In *Oeuvres* (eds Garnier & Flammarion), T.2., Paris.

Spinoza B. (1966) Traité politique. In *Oeuvres* (eds Garnier & Flammarion), T.4., Paris.

Weber, J. (1992) Environnement: développement et propriété. Une approche épisté-mologique. In *Gestion de l'environnement. Ethique et société.* pp. 268–301. Fidès, Québec.

Note

In the summer of 1994, three television transmissions in Italy were devoted to presenting the gravity of the situation facing the swordfish fishermen and the possible bio-technical and social solutions, based partly on the results of the successful experiment with bell buoys. But the conservation organizations – the Worldwide Fund for Nature, Greenpeace and some 13 others – have declined to join in the debate. Stranger still is the continuing silence of the European Commission, well acquainted with the results of the research yet unwilling to consider alternative solutions to the virtual closure of the fishery.

Editors' Note

Drift net fishing is a confused and contentious issue. The EU's ban on drift nets over 2.5 km was in compliance with the UN's moratorium. According to some sources (*European Report*, No 2045), scientific evidence in the Mediterranean suggests that only 20% of the catch is made up of the target species, the swordfish. The Council of Ministers had agreed to postpone a decision on phasing out all drift-net fishing, including nets under 2.5 km, pending further scientific evidence. Although such evidence was presented in February 1995, the Ministers have yet to take a decision. A view expressed by COREPER on the scientific evidence suggested that a ban on driftnets might reduce effort in the short term, but with severe socio-economic consequences, while in the long term, total effort was unlikely to be affected by such a ban (*European Report*, No 2049). A Commission proposal to ban their use for tuna, salmon and swordfish fishing in EU waters cannot so far achieve the required qualified majority support (*European Report*, No 2055). The apparent reluctance of the Council of Ministers to impose further restrictions is in marked contrast to the European Parliament's approval of a virtual ban on all drift-net fishing from the end of 1994.

Reference

European Report (1995), European Information Service, Brussels.

Chapter 11
Management and Practice in the Small-scale Inshore Fisheries of the French Mediterranean

ANNIE HÉLÈNE DUFOUR

Introduction

To place the fisheries of southern France within the overall national context is likely to prove a difficult exercise, so greatly do they differ from the national situation whether by reference to the fishing methods, the way production is organized and the fishing grounds managed, the size of the crews, the rhythms of work and so on. By contrast, relating them to their other frame of reference – the Mediterranean – is very much easier because there are many more features which the southern French fisheries share with their Mediterranean neighbours. These feature are, of course, concerned with the oceanographic characteristics of a semi-enclosed sea, with a narrow continental shelf, relatively high water temperatures, and also with the great diversity of technical features common to the countries of the Mediterranean (Italy, Spain, Morocco, Algeria and Tunisia, in particular) though in differing proportions. Moreover, when referring particularly to the western part of the basin, the types of traditional, local trade organizations such as the *cofradías* in Spain and the conciliation and regulatory boards (*prud'homies*) in the French Mediterranean, have much in common. Both have made strong imprints on the modes of management of the fishing grounds and on the social life of the fishing communities.

What characterizes the fisheries in the French Mediterranean – and more particularly the situation in Provence – is the craftsmanship, the dispersion of fishing areas, the small size of the fishing fleet in each of the harbours and the multi-specificity of the skippers, with each skipper able to conduct several different fishing activities. All this contributes to an extreme variety of practice, mainly due to the fact that these activities are, for the most part, undertaken by what the industry refers to as 'small crafts' *(les petits métiers)*. This term properly refers to fixed methods (*les arts fixés*) or the gear and tackle secured to the sea bed by some form of ballast and which thus remain stationary once they have been set down on the bottom of the sea. Such gear include nets, lines, long multi-hooked lines (*palangres*) and pots deployed by fishermen operating from small boats in rather shallow waters – rarely more than 50 fathoms deep – and generally within sight of the coast. These *petits métiers* exploit a very narrow littoral zone, as the continental shelf is greatly reduced in extent and scarcely exists in places; it is on this narrow platform that most of the marine fauna is concentrated. Moreover, the temperature of the deeper waters in the Mediterranean at 12°C below 100 fathoms, compared to 3–4°C in the Atlantic Ocean, is inimical to some of the species exploited on the Atlantic continental shelf.

The extraordinary diversity of the technical means, and the mutli-specialization of the men who use them, underlines the importance of optimizing the management of the narrow coastal zone and the relatively restricted resources that it contains. In this respect, the harbours of the Golfe du Lion – such as Port Vendres, Sète, Marseille – should be distinguished from those located to the east of Marseille. Because of its much wider continental shelf, harbours on the Golfe du Lion shelter most of the trawlers, tuna boats, sardine boats and drifters. Marseille, lying at the junction of the two areas, combines characteristics of both the Provençal coast and the Golfe du Lion. In all, 92% of the large offshore fishing fleet is located in ports along the Golfe du Lion, with Marseille and Martigues accounting for 28%, whereas most of the fishing fleet to the east of Marseille – together with Marseille itself – comprises *les petits métiers*. The Languedoc coast has what Giovannoni (1987) termed 'a double-sided littoral' (see Fig. 11.1) – the outer one facing the sea and the inner one bounded by salt-water lagoons (*étangs*). The *petits métiers* are concentrated mainly in the lagoons, except for seine net fishing traditionally carried out along the beaches of the sea coast. Traditionally, local fishing in Languedoc has always taken place in the lagoons, whereas high seas fishing has principally been carried out by fishermen from other regions, including the Catalans, Italians and the French from Algeria or Tunisia (Giovannoni, 1987; 1988). To cut a rather long story short, in Provence – just as in Languedoc – it is always small scale fishing that makes up the essential part of the so-called 'traditional fishery'. It is mainly represented by *les arts fixés* and, in Languedoc, also by drifting gear, except for the *gangui* fishing method using towed gear (*les arts trainants*). *Gangui* fishing refers to seine nets and other gear towed vertically from the back of the boat for some distance before being hauled aboard the boat. This also takes place in shallow waters and within sight of the shore, that is within the three mile zone. As this system contravenes general fishing regulations in France, it requires a special licence and this also helps to account for its scarcity (Dufour, 1987).

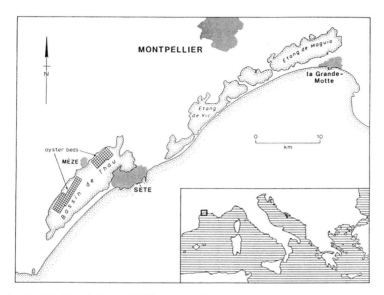

Fig. 11.1 The 'double-sided littoral' of the Languedoc coast.

The formation of local, private knowledge

Sound knowledge of the marine environment and, in particular, the configuration of the sea bed is an essential pre-condition for its exploitation (Dufour, 1990). Thus, the fishermen need to possess a global inner image, a true mental map, of the waters where they work. This is the fruit of experience and of the knowledge handed down from generation to generation both by word of mouth and by observation and gestures; when dealing with fishing, demonstrating and observing are the essential means of any form of training. But how is the mental map constructed? If the fisherman is to know the local sea bed as well as a gardener knows his garden, then knowing means mapping in one's mind and carrying an accurate image of the marine environment in one's head.

'One sees the bottom of the sea, as it were, without seeing it; one sort of images it.'

Observation of the sea bed – for the most part impossible to observe directly – requires some form of intermediary: the nets, long lines (*palangres*) and pots. Their behaviour on the sea bed and the response to currents and tides, together with the contents of the haul, reveal a good deal about the habitats of the sea bed and the marine fauna it shelters. Perception of the sea bed – a prelude to 'knowing' it – is thus achieved indirectly through the mediation of a particular fishing gear which, conversely, has been designed on the basis of that knowledge. Further, the comprehensive configuration of the sea bed from the shoreline to the break in slope that marks the outer edge of the continental shelf – beyond which the *petits métiers* do not operate – is the object of collective knowledge and representation. The relief, vegetation and detailed characteristics of the ground form the basis of the spatial structuring of the fishing grounds. From such precise knowledge, together with practical information such as the types of fishing and distance from the shore, another crucial spatial dimension is added to the mental map; this time the knowledge is more functional as it relates to the selection of the most favourable fishing grounds.

Knowledge of the sea bed is, of course, combined with other kinds of knowledge concerning submarine vegetation, climates and currents to allow the fisherman to determine the proper manner for setting the net (*caler*), positioning the net according to a particular orientation. Implicit in all forms of local knowledge is an understanding of the behaviour, habits and preferences of the fish, which will alter throughout the annual cycle; once again, such knowledge results from patient observation and the handing down of information from one generation to the next. It is this 'science', accumulated day by day, which enables the fisherman to anticipate his chances of making the expected catch in such and such a place, during such and such weather and according to such and such a season.

Within the common territory, therefore, each fisherman has his own 'corner' – that area of sea which he has personally 'discovered' or which has been passed down to him by an elder, either his father or fishing boat master.

Local management institutions: the *prud'homies*

The exploitation of these individual fishing areas obeys a system of customary law based on secrecy and respect for the areas 'appropriated' through private knowledge, even where they happen to be widely known. Likewise, the transmission of knowledge follows the same pattern; by the end of his working life, an old fisherman will have passed the knowledge down to his son or the heir of his choice. This distinguishes the individual or 'private' fishing areas from the fishing 'posts' (*les postes*), that is the fishing areas that form the collective property and whose management follows an overall set of local regulations officially laid down by the fishermen's collective jurisdiction (*prud'homies*).

The *prud'homies*, a system of collective management covering a recognized local territory, form a very important and distinctive feature of the organization of the fishery in Mediterranean France. Heir to the 'brotherhoods' (*confréries*) and trade guilds (*corporations*) of the *ancien régime*, and far from being abolished during the French Revolution like the other guilds, the *prud'homies* were confirmed in their powers by the Constituent Assembly and even encouraged to develop in those harbours where they had not previously existed. From that period onwards, the Mediterranean coast progressively witnessed *prud'homies* being installed with a status similar to that of Marseille which, along with Collioures, was one of the very first to be recorded (Motais, 1981). Thus, any individual harbour – or almost any, as some regroupings can be found – has developed the institution to manage, regulate and supervise its fishing territory. Its limits, demarcated by decree, are determined by precise alignments, real borders – just like those of parishes and countries – beyond which regulations and customs may vary. This maritime territory, thus individualized and protected, is under the responsibility of the *prud'homies*, the elected fishermen's representatives.

The *prud'homies* are run as democratic institutions; all new projects and possible modifications concerning the fishing activities are voted in by a majority, after the members – that is, all the fishing boat masters – have been called into session. They are also endowed with juridical functions which enable them to settle differences between fishermen and any other offences committed within the maritime territories of which they are in charge. On similar grounds, the *prud'homies* also act as local courts of justice. Not all that long ago, the fishermen used to wear court dress – black cap and gown – to administer justice; today the dress is worn only for public ceremonies like St Peter's Day.

The *prud'homies* are also granted competence, within French fishing regulations, which allows them to share out the rights of exploitation of the sea among all fishermen. This is the aim of the Prud'homal Regulations (*Réglements Prud'homaux*) which, as they are enacted by each community of locally registered, practising master fishermen, can accurately fit the local conditions and also be quickly adapted to meet changing needs. Generally speaking, those regulations based on long experience of exploiting familiar fishing grounds 'appropriated' by local fishermen, have as their aims:

- a fair allotment of resources among fishermen from the same *prud'homie*, which is done by way of a periodic drawing of fishing posts (*postes de pêche*) for example[1];
- preserving the territory from intrusions, or at least controlling such intrusions – a fisherman who is a stranger to the *prud'homie* must first introduce himself to the *prud'homie* in order to practise certain types of fishing within its waters and he must, of course, respect the rules regulating the *prud'homie* that receives him;
- managing the resources through the regulation of times for the setting of nets, the size of the mesh, seasonal suspensions of fishing activities on spawning grounds...

Moreover, *prud'homies* undertake a number of other functions indispensable to fishing activities within the local setting. Their quarters are often used as a storehouse for part of the fishermen's equipment and formerly for the *péirou*, the cauldron or tank used for dyeing the nets. In some harbours, the supply of fuel and gears, the renting or the use and pooling of a dry dock also depends on the *prud'homies*, which will normally perform functions similar to those of a co-operative society or which will complement the activities of the co-operative society if one exists.

They had long played the role of friendly societies for fishermen before the advent of welfare legislation which today protects the fishermen and their families (relief funds for fishermen's widows, sickness benefits, bereavement contributions, temporary care of orphans, etc.). In fact, the *prud'homies* still perform this role today, whenever a tragedy such as a shipwreck occurs.

Finally, on top of all this, *prud'homies* also perform certain social functions. Their quarters, which are almost always situated at the heart of the harbour, are places where fishermen can meet regularly either to debate matters directly related to their trade or, less commonly, activities linked to it. Thereby, they have been – and many still are – the male centre for social relationships, occupying a space akin to both Mediterranean 'men's houses' and the *Provençal cercles*.

The long established anchoring of the *prud'homies*, their democratic features, the roles they play in the management of marine resources and their capacity to serve as mediators between state administration and local trade concerns, have contributed to shaping them as original and valued instruments in the practice and organization of Mediterranean fisheries. Because of their continuing adaptability, they fit particularly well the changing conditions of the fragile inshore fishing environment consisting of a narrow coastal strip of sea covering the Mediterranean shelf.

References

Dufour, A.H. (1987) Poser, traîner: deux façons de concevoir la pêche et l'espace. *Bulletin d' Écologie Humaine*, **5** (1), 23–45.

Dufour, A.H. (1990) Leggere e gestire i fondi marini. Due aspetti complementari della pesca nel litoral della Provenza. In *La Cultura del Mare. La Ricerca Folklorica*, **21**, 51–5.

Dufour, A.H. (1993) Les pêches traditionnelles, l'exemple Méditerranéen. In *Le Patrimoine Maritime et Fluvial: Actes du Colloque Estuaire 92*, Nantes, April 1992, pp. 192–7.

Giovannoni, V. (1987) *Des Jardiniers de l'Eau. Genèse d'une Culture*. Université de Montpellier, rapport dactylographié.

Giovannoni, V. (1988) *Le Mourre Blanc. Du Technique au Social*. Université de Provence, rapport dactylographié.

Motais, M. (1981) *La Pêche Française en Méditerranée*. Mémoire pour le DESS de droit maritime, Aix-Marseille.

Note

(1)   The drawing of lots for 'posts' on the fishing grounds – a common practice in the Mediterranean *prud'homies* – is the most highly developed expression of the concept of an economic activity based on equality and community spirit. It

impregnates all prud'homal texts. Here is one variant from Var, in eastern Provence.

The 'posts' are fishing areas long known as passing places for fish; recorded in the collective memory, there are a limited number in each *prud'homie*. Contrary to other fishing locations which are known to and appropriated by individual fishermen, with the tacit consent of other fishermen, the 'posts' represent collective property, openly exploited according to a written code and regulations meticulously set down by the fishing communities. If all 'posts' are productive, they are not equally so; therefore, the drawing of lots creates equal chances for all, while preventing personal appropriation of a property considered as collective by the very principle of drawing lots. Once provided with a 'post', which chance has given him, the fisherman becomes its proprietor for 24 hours but under certain strict conditions: (i) he must set his nets before sunset, otherwise the 'post' will be given back to the community and be available to the first fisherman to arrive at the site; (ii) then he must lift his net the following morning an hour after sunrise for the next 'proprietor' to take his turn. Indeed, from the same drawing of lots, a rotation is set in motion for all the 'posts' contained in the draw: the fisherman who has drawn the last 'post' on the list is next granted the first one, while the other fishermen shift down one rank in numerical order.

So, at the end of the rotation, each fisherman who has taken part in the draw will have exploited each of the 'posts' included in the draw. Once the cycle is over, a new allotment may take place in which all the 'posts' are redrawn thus giving everyone a new chance. The same operation will be repeated by the same modalities just as long as the fishermen, at the end of each cycle, reapply for the lottery – until the end of the season for 'post' fishing.

This procedure secures a periodic redistribution of the most highly rated fishing grounds – the others are not included in the allotment. In an activity where resources can fluctuate quite markedly, the rapid rhythm of rotation allows even the most unlucky ones not to suffer too much from a bad 'post', just as the luckiest may not benefit too greatly from their good fortune. Thus the equality of chances among all the participants is seen to comply with a double warranty: that of fortune, which decides the initial allocations, and that of the community, which re-arranges them so as to give fortune a fairer sense of direction.

Part IV
Local and Regional Dimensions of Fisheries Policy

Chapter 12
Fishermen Households and Fishing Communities in Greece: a Case Study of Nea Michaniona

KATIA FRANGOUDES

Social history intends to identify and explain social changes by focusing interest on the observation of social groups. The evolution of practices and organizations specific to a group is one of the key research areas, and case studies are a major source of information. Fishermen and their households are the target social group of this particular contribution, in which fishermen are understood as all those individuals who make their living from predation on living aquatic resources. They share several characteristics in common: their activity depends largely upon natural variability (abundance, weather); in most cases, the use of marine space and its natural resources occurs under common property regimes; and the specificities of their technical and geographical areas of work further contributes to the strong identity of the group. A social group, like fishermen, is however not only defined by reference to its specificities; it is also, on the one hand, part of the global social environment and its history and, on the other, heterogeneous in character. Within this perspective, we look for facts related to the history of individuals, local communities and any other sub-groups defined, for example, by use of a particular fishing gear, to form evidence for the understanding of social changes at the group level. During recent decades, European fisheries have experienced major institutional, technical, market and other social changes. As a contribution to the understanding of such changes encountered by fishermen, the interim results of an on-going case study in a Greek fishing community are presented here.

Michaniona: a refugee fishing community

Michaniona is a wealthy town (6000 inhabitants), located 30 km south of Thessaloniki on the east coast of the Thermaikos Gulf, one of the most productive fishing areas of the eastern Mediterranean. Michaniona is a fishing community in the sense that fishing is the core activity of the local economy. Practically all aspects of the local economic and social life of the town depend upon the fishing activity which directly employs 60% of the working population and lands 12 000 t of fish every year. Fifty-four trawlers and seventeen purse seiners form the local component of the medium-range fisheries and about one hundred small boats are registered in the coastal fisheries[1]. Today the main target species for purse seiners is the anchovy, while demersal species form the basis of the diversified landings from the trawlers.

The fishermen in Michaniona have had to deal with various endogenous and exogenous events and developments. Most of the families in Michaniona originally come from two villages, one on the Black Sea and the other in Asia Minor. They were Greek refugees who came to the Greek mainland after being expelled by the Turkish Government in 1922. They chose this area, almost unused at the time, to build their village. The main reason for this choice was that most of the families were involved in fishing before migrating and had already fished the Thermaikos Gulf and around the Aegean islands (see Fig. 12.1). The village was named Nea Michaniona, New Michaniona, from the name of the original village on the Black Sea where the majority of the families originated. In their new settlement, they naturally turned to fishing. Although the conditions of the migration are subject to some controversy, it seems that the biggest fishing boats were used in the escape from Turkey. So, not only did these families come with their fishing knowledge, but also with their boats and gear. Since that time, fishing has remained the major economic activity.

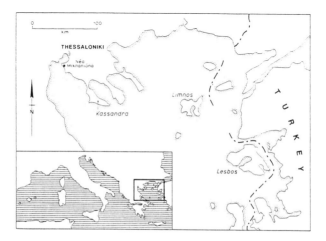

Fig. 12.1   The northern Aegean Sea.

Some indication of the instability of the fisheries, caused largely by external market conditions, can be gained from a brief review of the rise and fall in popularity of particular species. In the beginning, the main fishing technique was the purse seine, for the people coming from the Black Sea, and purse seines and trawls for the others. As the former comprised the majority in the new settlement, purse seine fishing was the dominant activity and the main target species were sardine, bonito, striped mullet, sea bream, sole and whiting. Sardines were supplied to domestic salting industries; anchovy and tuna were at that time discarded as no market could be found for them. Commercial anchovy fishing started up in the 1960s and it eventually became a target species for the purse seines during the 1980s. Part of the catch is salted and the rest destined for the fresh fish markets in Greece and beyond. Six purse seine vessels are now equipped with nets for tuna fishing following a rise in demand from Japanese importers. For a few years, shrimp and Norway lobster also became a major source of income as marketing opportunities developed. After the demand for sardines fell, some years elapsed before the small pelagic species again became recognized as a valuable resource; many fishermen changed from purse seining to trawling and the trawl became the dominant gear for the medium-range fisheries. Small-scale fisheries have always formed an important part of the local fishing activity and, following a period of progressive decline, they have recently been reactivated with the discovery in the late 1980s of the economic potential of *Venus verrucosa* (wart venus). A large part of total production goes for export to Italy, a market with a strong demand for sea foods that its own domestic fisheries can only partly satisfy. Domestic demand in Greece is also rising but, for many species, cannot compete in prices with the Italian market and a recent decline in living standards has affected the growth in demand for sea products.

Up to the start of World War II, the fishing fleet was comprised of sailing boats. Motorization occurred during and after the war and, since the end of the 1960s, various forms of electronic fish detection apparatus have been installed on board and deck operations have been extensively mechanized. Recent modernization has been accomplished in part with the aid of EU grants following Greece's accession in 1986. Improved efficiency in the hunting of the fish, in reducing the number of crew and in limiting the possible causes of accidents are the main motivations for technical investment on the boats. Today, the fishermen will exploit all the possibilities offered by modern technology, especially on vessels in the 18–25 m range, in pursuit of greater operating efficiency.

Various forms of restrictions have been imposed on the local fisheries in order to restrain fishing effort and conserve fish stocks. Trawling is forbidden during the summer months, while purse seining does not operate in the winter. Expansion of the medium range fisheries has been controlled by a licensing system since 1966 (Ministry of the Government Presidency, 1970) and no new licences have been available since 1991, except where an equivalent effort has been withdrawn from the fleet. In 1978 the northern part of Thermaikos Gulf was closed 'temporarily' to trawling in order to avoid conflicts over benthic and demersal species with the coastal gill-net fishermen from fishing communities located around the bay. The area for trawling operations has thus been reduced and despite a call by the trawl fishermen to reopen trawl fishing in the northern Thermaikos Gulf, it remains closed.

Changes to working conditions

Technical innovation and changes in social regulation have profoundly modified the working conditions of some fishermen. The purse seines used to employ a crew of 30 until 1954, when the first social insurance system was applied to fishing boat crews. The mechanization of net operations has also contributed to the reducing of crew size. Today, there are only 12 to 14 crew on the purse seiners. Similarly, the trawlers are today operated by only five persons, where formerly they employed 10. Most of the medium-range fishing boats are jointly owned by brothers or relatives. The first generation of refugees, who owned the large boats, generally hired both skipper and crew and did not go to sea themselves. But, by the time the crew size had been reduced, the second generation of boat owners had started to go to sea themselves.

The family business character of the trawling operations in particular has been reinforced. In a few extreme cases, the trawlers are now operated only by the owners and members of their families. But the continuity of this form of organization based on family succession and on-the-job training is under threat. Associations of relatives set up to jointly operate a boat will tend to increase as no new licences are being issued and while fishing continues to represent a good opportunity in the present conditions of the Greek economy. But when three or four relatives jointly operate a boat where no more than five crew are required, there is little or no room for them to train their children for work on the trawler. The transmission of know-how and property becomes an increasingly complicated issue and is already causing some internal family conflicts. The situations will become even more difficult if the number of licences is actually reduced in line with the objectives of the Common Fisheries Policy.

To reduce labour costs, boat owners have called on foreign contract labour, mainly Egyptians originating from fishing villages, and in most of the main fishing ports in Greece they account for the majority of the labour employed on the purse seiners and trawlers. At the outset, the contract labour was paid a salary free of any form of social insurance (Frangoudes, 1993). But in Michaniona, strong opposition from the union of fisheries workers forced the boat owners to hire foreign workers under the same conditions as for Greek labour, including the share system and payment of social insurance. As a consequence, 90% of the labour on Michaniona fishing boats is still hired locally. The recent inflow of Greek refugees from Russia offers further opportunities for hiring cheap labour. But very few have been brought to work in Michaniona, on the grounds that they have no relevant experience. In this respect the local community is still sufficiently well structured to be able to resist competition on the labour market. Working on a fishing boat is still very positively evaluated and the crews' union continues to play a central role in maintaining the positive image. A convention exists between the crews' union and the Government, stipulating that to employ a new crew member requires the agreement from the crews' union. The union participates in the calculation of the shares for all boats and it plays a particularly important role during the few months of overlap in spring and autumn, when trawlers and purse seiners operate on the fishing grounds together. At this time, there is a shortage of skilled crew members and the union can regulate the labour market by authorizing the employment of a limited number of foreign workers.

There is a strong resistance to the introduction of certain new fishing techniques, such as pelagic trawling and the dredging for wart venus. In the small pelagic fisheries, the introduction of pelagic trawls would mean a sharp increase in fishing capacity compared to the purse seine, for an equivalent number of boats. In most areas of the world where pelagic trawlers have competed with purse seiners, the latter have progressively disappeared. At present there is no direct competition for the resource between bottom trawlers and purse seiners engaged in light fishing at night; and pelagic trawling is forbidden by law throughout Greece, as is also the case in Spain (Frangoudes, 1995). While dredging is authorized for oyster fishing, a rule prohibiting dredging for wart venus is actively supported by the majority of the 200 wart venus fishermen operating in the gulf, of whom 100 are from Michaniona. Wart venus are collected at depths of 8–10 metres by diving with the assistance of on-board compressed air. Each boat normally carries four persons; three divers and one person to supervise the operation, as required by law. Although it is not known whether the stock can support the existing level of fishing effort, it is certain that the introduction of dredging would require a serious decrease in the number of boats and fishermen, if it were to remain sustainable.

Women also used to be more involved in fishing related work. Most were engaged in preparing and maintaining the nets. But, with the introduction of nylon nets, there is less work required although to have it done outside the local community would be very expensive. Today, there is only one surviving net workshop, employing eight women; most of those employed are getting old and the hard working conditions make it unattractive to others. Some wives have taken over responsibility for administrative and financial work related to the fishing enterprise – activities which formerly were undertaken by paid employees.

Regulating resource, space and market competition

Instances of resource, space or market competition are numerous. Collective resistance against an increase in competition, as expressed above, and the settlement of disputes by face-to-face negotiation are the two main answers. Meetings around the bar room table, together with the involvement of a few professional organizations, provide the opportunities for collective resistance and face-to-face negotiation. But competition sometimes turns into conflict situations. Competition regulation and the settlement of conflicts, among gears or between fishing communities and with middlemen, have many dimensions.

Local initiative and central rule

For a better understanding of the relations between central and local power, we must first recall that national unity in Greece is both recent and characterized by a strong state centralism. There is no long tradition of decentralized institutions for fisheries management to compare with the *prud'homies* in France or the *cofradías* in Spain, both born in the Middle Ages, nor even with the corporatist professional organizations built up in France since World War II (Weber, 1993). When a dispute among individuals or groups cannot be settled by face-to-face negotiation, it is left to arbitration in Athens. It is a one-way, two-step process. The local level first asks the state for a ruling through the administrative channels – generally the local fisheries inspectorate. Then, when the request reaches the upper level, fishermen's leaders enter into direct negotiations with the central authorities in Athens. These leaders are either representatives of fishermen's organizations or simply well-known and respected elder fishermen who maintain a network of close personal relations with the centre through political or other channels (Frangoudes, 1993). In some cases, fishermen's groups may pay a counsel to defend their interests against the fishermen of a neighbouring village and obtain a favourable decision from the central authority. The practice of engaging a counsel is very common in Greece; boat owners' organizations and crews' unions, for example, regularly use the services of counsels. Political links are also used for the same purpose. The result is a complicated set of interventions and sometimes contradictory laws, decrees and regulations. All means of regulation – such as gear restrictions, time or area closures – are used to regulate the fisheries from the national to the local level, under the aegis of the central authority. The length of the formal process sometimes means that the official decision is arrived at some time after informal, direct negotiations have reached a solution and is thus redundant. By contrast, in other cases the speed of official response can be very rapid: recently the demand for a two-day extension of the purse seine fishing season was granted within two days. As elsewhere, a 'no decision' response may well be adopted, particularly where the issue is politically very sensitive.

Considering the limited means of control by the fisheries administration, effective implementation is left in most cases to self-control by the fishermen. The local administration will act against infringements when there is strong demand from other fishermen. Thus social control plays a very important role in the observance of regulations. But when the infringement is no longer a cause for strong opposition, the regulation may simply be ignored and only reactivated later should the need arise. The appearance of European institutions, as regulation and grant providers, considerably disturbs the traditional system for generating and implementing rules. Local representatives and the elder statesmen of the fishing industry find it very difficult to build close links with the new central institutions. Nor are they encouraged to do so by the national authorities, who use the 'Brussels decision' argument for their own convenience with little means of control therefore left to the fishermen's representatives.

Time and area closures to limit fishing effort and fishing gear interaction

Time and area closures are a very common regulatory tool for Greek fisheries. Many such regulations find justification today in resource conservation but the original motivation and conditions for a good social acceptance of these limitations varies from one case to another. They are sometimes difficult to elucidate either because no record of the facts is kept or because the issue had been so conflictual that no one wants to talk about it. As an example, a four month closure of trawling activities in summer has been legally recognized since 1954 and seems to have been practised for an even longer time. Possibly, it was originally justified on religious grounds (a summer fast), for marketing reasons or even as a means to limit competition. Summer is the main season of activity for the purse seine. The traditional process of fishermen's initiative/state decision/local control of implementation has in recent decades produced other restrictive regulations. Hundreds of cases of time and area closures are recorded (Adrianou, 1987). For example, three winter months of closure for purse seining have been applied since 1991 and a four month ban on clam fishing was introduced in the north east of Thessaloniki Bay in 1994. Very little biological research is conducted in justification of these decisions or, where such studies do exist, the results are generally not available to the fishermen who seek to recommend new regulations to the central authority. Although they may afterwards be justified from a biological or ecological point of view, time or area closure decisions are more usually the result of more or less conflictual negotiation processes among the fishermen, either within a single gear group or among different gears. This has most recently been observed in the case of venus fishing. It also occurred in the very serious conflict that led to the decision to close the north of the Thermaikos Gulf for four months. The fishermen use their practical knowledge to formulate the issues of a conflictual situation and, when agreed by the majority, they request that all possible means are taken to avoid illegal fishing. For example, in the case of lengthy closure periods, the gears are sealed by the administration. Only afterwards is the decision justified on the grounds that the closure period was universally recognized as the spawning period.

Boat owners react in different ways to these restrictions. Those who benefit from both trawling and purse seine licences simply turn their boat over to purse seining in the summer. The length of the trawling closure makes it worthwhile reorganizing the boat and the operation. But relatively few boats have this opportunity for flexibility. Before venus fishing was closed for the summer, trawl owners who had inherited a coastal boat would go diving. Some consider that they do not need to go fishing during the summer months, although the idea of seasonal work in other sectors of the economy is not well developed.

From the example of the trawl closure, we can see that the conclusion of a negotiation involves many factors – amongst others, the ability to build a strong interest group within the fishery or at a village level. There is also room for clientelism using the fishermen or the political representatives. An example is the special permission given to a few trawlers from Michaniona in 1994 to operate in international waters during the trawling closure period. According to one local source, this decision almost brought about 'civil war'. Other fishermen demanded that they be given the same authorization – some only on principle, knowing full well that they would not go anyway. The merchants also entered the debate; using the argument of the spawning period, they argued that the fish were too small for the market. In actual fact, they were afraid that landing fresh fish during this period would prevent them from selling the stocks of frozen fish they had built up during the previous months and which are sold at very high prices because of the general closure. In the end, the special permission was withdrawn, but the consensus over the trawling closure season had been put at risk.

One of the consequences of the EU's policy on grant aid has been to increase competition among the fishermen. Modernization and construction of fishing boats directly increases the pressure on local resources, which are also threatened by pollution and access limitations elsewhere in the Gulf, including the complete closure of the northern part of the Gulf since 1978. Grants have also provided an incentive to borrow money from the banks. The burden of credit on the livelihood of some fishermen households is such that trawlers work almost permanently during the open season. To limit direct competition on the local fishing grounds, there is a tendency towards a geographical extension of the fishing activity. Some boats, for example, fish as a group off some of the Aegean islands, thus lifting part of the pressure on the local stocks; one boat makes the return journey to the nearest harbour on the mainland to land the catches and to allow control of landings and share calculations by the crews' union. Seen from Michaniona harbour, this is a good thing: competition is reduced and local employment and market supply are guaranteed. But little information is available on the competition on the new fishing grounds. One major consequence of this development affects the role of women within the household. While the fishing trips previously were never longer than 24 hours, some fishermen are now away from home for three months and their wives must therefore assume the responsibility as the head of the household during their prolonged absence.

The environmental question

Fishermen also express worries over environmental issues. At the head of the gulf, the urban concentration of Thessaloniki is a major source of pollution with the first water purification plant only now under construction. Some inflowing rivers, with polluting industries in their catchments, also cause concern to the fishermen. They observe the decline of Posidonae seaweed beds. The number of plastic bags collected in the trawl nets are sometimes so great that the minimum mesh size rule seems meaningless. With many different interests involved and little scientific research on the problem, the fishermen are unable to identify an effective locus for the expression of their concern. They sense themselves the victims of the problem but with no means of action to prevent it; their only opportunity for projecting their fears is to suggest that they be used to 'clean up the sea'.

Collective organizations and attempts to regulate marketing

The main organizations to represent the fishermen are the boat owners' co-operatives, originating from the 1950s and more in the nature of associations, as defined by Greek legislation, than true co-operatives. They are organized on a gear basis – the trawlers' cooperative, the purse seiners' cooperative and the coastal fisheries' co-operative – and their main function is to represent and defend the interests of their members. Although only a minority of the trawler owners are members of the co-operative, all the purse seine owners and most of the coastal units belong to their respective co-operative. During the 1980s, various attempts were made to organize their own marketing structures. The trawlers' co-operative started to operate as a producers' organization, but it broke down after two years of operation. The purse seiners similarly tried to organize a collective marketing system but conflicts arose among the participants. As an outcome of this last initiative, the majority of purse seine owners still maintain a collective marketing structure, managed by two relatives of the fishermen, which operates on the basis of a fixed margin. The coastal fisheries' co-operative started to market the venus on behalf of their members in 1993. Several other attempts at collective organization of marketing are recorded, all with very short lives. In 1990, the trawler owners not linked to the co-operative decided to market their production jointly; fishing areas were to be allotted to the individual fishermen and the proceeds of the sale of fish were to be shared equally. The arrangement broke down after only six weeks. It is the search for an alternative to the middleman that most often motivates such attempts to build up collective organizations for purposes other than representation. They seek to apply precise egalitarian principles that prove very difficult to implement in practice. Currently, the decision to build an auction hall in Michaniona is the subject of strong controversy. Up to now the fish has been landed in Michaniona and trucked to the auction hall in Thessaloniki. A new auction hall has been built in Michaniona but is not yet in operation, as the merchants refuse to move to the new hall and instead are demanding the construction of another hall on the opposite side of the bay.

Conclusions

In many respects, Michaniona's fishing community appears to be very dynamic and to be able to adapt rapidly to change. It is not the only case in the Mediterranean where fishing community dynamism is partly based on a 'refugee' identity (Frangoudes, 1995). The dynamism applies on many levels: changes in demand and in technology are two major causes of the evolution. The first explains the important role of middlemen as transmitters of information about the market. But it is also the ground for resentment against those who seek to impose their own law. As demand has become more and more sensitive to market competition, the fishermen have tried to gain independence from the middlemen. The initiative to self-organize on the marketing side is very representative of the group approach to collective organization. But the search for equity, that seems to play such an important role in defining the social recognition of fishing limitation decisions, does not seem to work towards making viable marketing arrangements. Despite this, the past decade has seen many renewed attempts in this direction, always with a lot of enthusiasm. Dynamism in the introduction of technological change is balanced by the refusal to admit certain fishing techniques that would lead to a rapid decrease in the number of fishing units and in employment. As a result, fishing still employs many people locally. Very strong institutional arrangements to regulate the labour market have played a central role in this local dynamism.

The process of framing rules and the motivation and condition of social acceptance, appears a subtle mix of professional skills, face-to-face discussion in bars or in private, local and central administrative involvement, political and personal links, the intervention of counsels and law suits to try to regulate competition while balancing interests. Living with the fishermen gives an idea of the difficulty in implementing such regulations. Causes of competition and conflict are numerous. There is not a week without a case for discussion – a net destroyed by a boat, administrative or implicit rules broken by a fisherman, appeals to change rules previously agreed upon.

Here, we have highlighted only competition among fishermen, but conflicts with other users of maritime space and resources are also a source of concern – environmental degradation, leisure fishing *inter alia*. The increasing role of European institutions inevitably alters the rules of the game. The links and channels that were traditionally used to produce fishing regulations have to be redesigned to cope with the new conditions. The way fishermen's representatives and the administration build new links and how they adapt the decision-making process is a major institutional change which needs to be studied in greater detail.

The strength of time and area closures, compared to other possible means of regulation, suggests that they enjoy the highest degree of social acceptance. It would be interesting to understand better why they are so extensively used, rather than the technical measures. Less contestable, they may be easier to implement. Increasing limitations applied to fishing, and difficulties in maintaining incomes, forces fishermen into various individual and collective strategies.

Geographical extension of the fishing gear is the most significant individual answer. Gear flexibility is limited under present management conditions. Demands for the removal of limitations initially designed to be 'temporary' are growing. But the grounds for the conflicts that originally motivated them still exist and the administration, as well as some fishermen, are reluctant to take the risk.

The conditions of know-how and property transmission are also changing. Fishing, as a boat owner or crew member, is still socially highly valued. But a trend toward the reduction in the number of fishing units and the extension of the joint family operation will make it difficult to ensure succession and a permanent involvement in the fisheries. The problem is for the coming generation. Already it is difficult for some families to ensure on-the-job training for all their sons. Household and community life will be much affected if the practice of long fishing trips to distant places develops further.

References

Adrianou, N. (1987) *Fishing Techniques and Fishing Gears*. Le Pirée, Greece.

Frangoudes, K. (1993) *Le Rôle des Organizations Professionnelles dans la Gestion des Pêches en Méditerranée, Étude de Cas Concernant la Gréce*. Rapport pour la DG XIV de la CCE.

Frangoudes, K. (1995) Case studies in the Mediterranean Greece and France. Annex to first intermediate report. In *Management of Renewable Resources: Institution, Regional Differences and Conflict Avoidance Related to Environmental*

Policies and Illustrated by Marine Resource Management. EC-DG XII, Environmental Research Programme, contract EV5V-CT94-0386.

Ministry of the Government Presidency (1970) *Fisheries Code.* Decree 420/1970, Ethniko Typographico.

Weber, J. (1993) *Le Rôle des Organizations Professionnelles dans la Gestion des Pêches en Méditerranéedes: Synthèse.* ASCA–CCE DG XIV, contract XIV-12/MED/91/010.

Note

(1)   Atlantic, medium-range and coastal fisheries are the administrative categories used in Greece to classify fishing boats. Atlantic shrimp fisheries have almost disappeared. Trawlers and purse seines more than 14 m long are recorded as medium-range fisheries. Other gears are considered as coastal irrespective of their geographical range or the length of boat. For example, swordfish longliners are classed as coastal, despite the fact that they exceed 14 m and operate throughout the international waters of the Mediterranean.

Chapter 13
The Breton Fishing Crisis in the 1990s: Local Society in the Throes of Enforced Change

GENEVIÈVE DELBOS and GÉRARD PRÉMEL

Introduction

In 1804, a report from one of the first prefects in Brittany appointed after the proclamation of the Republic refers to 'the deep crisis that irreparably hits the Breton fishing industry that, with its 2000 boats, was so flourishing before the Revolution' (Le Bihan, 1958). Two hundred years later, the same remark can be made in more or less identical words, even to the number of boats involved. One needs only to replace the Revolution with the European Market or perhaps GATT.

This comparison helps to underline two points, first the durability of both the social environment linked to the fishing activity and the approximate size of the means of production; and, secondly, the need to reconsider the notion of crisis. Is it not conceivable that the 1994 crisis is neither more nor less irreparable than that of 1804? And during the two intervening centuries, just how many crises have been overcome?

In Brittany, the exploitation of marine resources, fishing and aquaculture is generated by complex and polymorphous societies with multiple but real dynamisms acting on several spatial and temporal scales. Starting with these social and cultural realities, could not the policy making institutions interpret such societies in a wholly different light, seeing them not as populations in need of outside assistance but rather as actors directly involved in the search for political and economic solutions?

These are the questions we wish to explore, while striving to distance our analysis from the climate of passion and tension that has so impregnated the Breton fishing environment over the past two years.

The Breton fishing industry in context

In order to characterize fishing-dependent regions, we can simply take the number of employees as an indicator. Today, for example, Brittany has 7000 fishermen, accounting for 45% of the total number of France (Didou, 1994). However, these figures only make sense when placed in a longer term perspective: in 1971, Brittany had 12 400 fishermen but accounted for only around 25% of the national total. This 20 year period thus emerges as a time of concentration and specialization of the industry in Brittany despite the significant decrease in the total number of fishermen.

The situation in Brittany may also be located in a European context. Two thousand fishermen is twice as many as the Netherlands, three times as many as in the former Federal Republic of Germany, only a little less than in Denmark but a mere tenth of the total number in Spain. Brittany is, in fact, the third most important fishing region in the European Union, after north east Spain and Scotland.

By adding the number of other jobs directly related to fishing and aquaculture, we can estimate the total number of employees as between 10 000 and 12 000 persons. The variation between the two estimates is attributable to the large seasonal variations in the number of employees during the year (see Table 13.1). Moreover, the activities of fishing and aquaculture together generate in the order of 30 000 and 40 000 shore-based jobs. All in all, therefore, we can count around 50 000 jobs related to sea-borne activities. By way of comparison, it is worth noting that farming in Brittany – the leading farming region in France – accounts for 110 000 jobs. The scale of the marine-based economy in Brittany can also be measured in terms of the harvest. The 2000 Breton fishing boats unload around 50% of the French catch (45% of fresh fish and 70% of shellfish).

Table 13.1   Numbers of fishermen employed in Brittany in 1993.

Size of vessel	Southern Brittany	Northern Brittany	Total
LHT < 12 m	1 828	1 391	3 219
From 12 to 16 m	906	365	1 271
From 16 to 25 m	1 995	683	2 678
From 25 to 38 m	634	46	680
Above 38 m	993	257	1 250
Others	776	281	1 057
Totals	**7 132**	**3 023**	**10 155**

Taking into account the relative importance of the number of employees and the scale of regional production, the most significant indicators of the recent crisis are related to the changes in the number of boats (see Fig. 13.1) and to the tonnage landed over the last six years (see Table 13.2).

Whilst acknowledging that the Breton crisis can be measured internally, the relative position of France in the global fisheries economy is also of relevance. While France ranks only nineteenth in terms of the global production of seafoods and Europe fourth, the EU is a leading world market for seafoods importing 60% of its needs. France, the

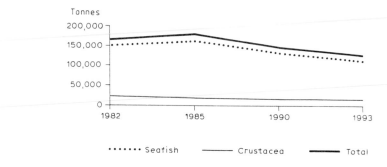

Fig. 13.1   Fish landings by Breton boats, 1982–1993.

Table 13.2 Development of the fishing fleet (1988–1993) in relation to activity.

Vessel characteristics		1988	1989	1990	1991	1992	1993
L < 12 m	Number	2 225	1 950	1 695	1 288	1 270	1 250
	Power	125 459	121 505	116 037	96 238	96 750	95 906
	Tonnage	11 015	10 239	8 500	6 988	6 980	6 871
12 ≤ L < 16 m	Number	373	378	359	328	323	320
	Power	68 042	71 500	70 502	66 053	65 549	64 558
	Tonnage	9 639	9 941	9 508	8 832	8 810	8 675
16 ≤ L < 25 m	Number	350	367	372	634	368	360
	Power	113 295	122 137	126 035	125 698	127 651	124 891
	Tonnage	19 902	21 393	21 963	22 201	22 969	22 619
25 ≤ L < 38 m	Number	86	96	79	62	54	54
	Power	46 218	47 609	45 002	37 222	321 169	321 169
	Tonnage	17 690	17 765	16 396	13 315	11 638	11 638
L ≥ 38 m	Number	69	69	69	65	60	59
	Power	123 575	123 722	130 330	128 636	120 904	119 800
	Tonnage	52 471	52 627	54 511	53 726	51 151	50 821
Total	**Number**	**3 102**	**2 850**	**2 574**	**2 107**	**2 075**	**2 043**
	Power	**476 589**	**486 473**	**485 906**	**453 847**	**443 023**	**437 324**
	Tonnage	**110 717**	**111 965**	**110 878**	**105 062**	**101 548**	**100 624**

second largest market in the Union, has the greatest external trade deficit in the EU for processed seafoods.

But all these quantitative criteria are inadequate to describe the far more complicated human reality. In particular they fail to explain the real levels of change which the coastal communities must confront. It was made clear in the introduction, that these areas which today are strongly associated with fishing and aquaculture have been influenced by a long history involving several different spatial and temporal scales. Over the centuries alternating periods of imbalance and equilibrium have helped to shape specific groups and communities, giving to each of them an identity and reference point by which they can meet the challenge of particular times and places and so renew their economic and social purpose. Today, however, that capacity for improvised response to global change is threatened by the increasing constraints of integration and external control to which all productive activities are now submitted.

By examining the nature and distribution of seafaring activities in each Breton coastal community, we can observe the way in which they reflect individual, family or collective strategies for survival in terms of crisis, based on alternations between different areas of marine activity. These can be summarized in relatively simple strategies:

- moving from one seafaring activity to another – from merchant navy to fishing or from fishing to shellfish farming – is quite a common practice in individual or family strategies based on 'diversion and return';
- exploiting the dynamics of industrial and small-scale fisheries: in a long standing region of emigration participation in deep-sea fishing (along with the navy) is very much in keeping with the colonial past and maintaining links with overseas countries. The 'Mauritanians' from Finistère who created the tradition of crow fishing off the coast of the former Spanish Sahara and Mauritania as a response to the crisis in coastal sardine fishing at the beginning of the century, are a good example of the dual dynamism (Boulard, 1991);

- within small-scale fishing: moving from one type of fishing to another and from one 'profession' to another with the definition of profession being related to the particular techniques used as well as to the species caught.

These combinations have contributed to the development of local specificities and traditions, the re-invigorated dynamism of actors who managed to deal successfully with the situation and the renewal of strong local cultures, each being known as 'Pays', for example 'Pays Bigouden', 'Pays de Lorient' and so on.

After all, communities of interwoven spaces and group stratifications could be defined in several different ways, each reflecting a different balance between the many different alternatives and leading to highly complex systems of identification. One is not simply a fisherman, but one also belongs, more or less, to the merchant marine; one is not only a deep-sea or an inshore fisherman, but one is either full-time or part-time and one practises trawling or line fishing, potting and/or net fishing. And at the same time, one may come from the Bigouden district, from Leonard, la Turballe or Quiberon, and so on. These subtleties of identity may significantly influence the composition of town councils in the fishing ports around the Breton coast, as well as the coast and departmental fishery communities. They also govern the rhetoric of crisis in the fishing industry.

In other words coastal Brittany comprises a series of overlapping yet distant entities, characterized by different historical influences, each with its own tempo and codes, its specific trade network, its own definition of membership and its own clan-like solidarity. These differences can promote rivalries between 'ports' or conflicts between the different 'professions', the one often overlapping with the other. By turning a point of difference into a potential source of envy, it seems that such rivalries may actually help to create the mainsprings of dynamism of the group, necessary to maintain it within the network of human interaction and ensure its continuing resistance to those forces which try to eliminate it from the economic and social scene. This is a fundamental point against which all administrative attempts at restructuring and standardization must bump, for their schemes are imposed from outside with insufficient knowledge and understanding of the reality.

This deep and complex sense of attachment to the resources of the sea has, until now, evolved in a way closely aligned to the fluctuations of both the natural or physical environment and the human or social environment. Reference to a marine tradition, to which Breton fishermen lay claim, is subject to constant redefinition. But it is this characteristic which almost certainly has allowed the Breton coastal communities to overcome the succession of 'irreparable' crises, which form the milestones of their individual histories. The situation is not unique to coastal Brittany; the strengthening of cultural identities through crisis also characterizes the history of Breton agriculture. According to some observers, it is the key to Brittany's broader regional dynamism.

'Out of the sea we make a living...'

A specific example, taken from the south coast of Brittany (See Fig. 13.2) and delimited by the peninsula of Quiberon and Le Croisic and the islands of Houat, Hoedic and Belle-Ile, provides a concrete illustration of those dynamics in action (Vince, 1966; Jorion, 1983; Mollo, 1992).

Fig. 13.2   The south coast of Brittany.

Until after World War II, fishing remains a complementary activity to farming or working on saltmarshes for most of the coastal communities of the Mor Bras. Fishermen from Finistère who travelled the coast in search of sardines and who had come to settle down with their wives and children, began to form a considerable colony of authentic fishermen; this trend continued until the early 1960s.

During the inter-war years, coastal fishing became a real profession carried out all year long. This move is particularly noticeable on the islands, where the maritime links with the mainland have been facilitated by motorization, and led to the retreat of agriculture. On Houat, this change is boosted by the emergence of a new opportunity: the prestigious Parisian restauranteur, Prunier, builds fishponds in order to receive regular supplies of lobsters and spider crabs which are plentiful in the locality. It is the beginning of a specialization that, through crisis and prosperity, will create the economic identity of the island.

In the aftermath of World War II, coastal fishing becomes exclusive to the periphery of the Mor Bras and down to Saint Nazaire. The sardine fishing boats that replaced the traditional fishing net by the *bolinche* have multiplied. Young country people who embark for the whole season (April to September) come to make up the crews of professional fishermen. But in summer a considerable number of boats revert to traditional occupations: the trawling for finfish (conger, hake, whiting, ray, sole and plaice) for daily consumption. The island of Houat is again distinguishable from the others: nephrops fishing is the main enterprise carried on all year long by crews of five to six men. They fish with separate pots for lobster, crabs and spider-crabs. Finally, small boats troll for bass, gilthead breams and mackerel. At the end of each fishing season, the produce is sold on the mainland to wholesale fish merchants.

The 1950s began with two disasters. In 1951, a violent storm devastates ports and boats; nephrops are no longer to be found. In 1956, a deep-sea port is built, but a year later more than the half of the 20 boats making up the local fleet are smaller than 20 t.

A radical change occurs: deep-sea gears for trawling appear, *bolinche* and other fishing techniques appear in the new port; 3 and then 4 m boats arrive. They specialize in sardine fishing from May to June/September with crews of 11 to 12 men. Some of the fishermen, owners of small boats and young people who do not yet possess their own means of fishing, gather to form the crews. They follow the migrating sardine down to the island of Yeu in the Vendee, remaining at sea for periods of eight to ten days except when fishing in the Mor Bras. Sardine fishing with *bolinche* boats always generates high returns. From now on, small boats using pots fish in lines of 10 to 30 boats according to their respective size. The less numerous line fishermen complete the fleet.

This equilibrium does not last for long. First of all, following the depletion of spider-crabs, the trade resulting from pot fishing collapses. Then, from the fishing season of 1964 onwards, catches of sardine dramatically decrease; canning industries which already import fish from Morocco quickly close down, one after the other. The golden age of the sardine for the inhabitants of Houat is over. But a new dynamism is forged by the crisis itself and the island becomes part of the regional economy.

Accumulated revenues allow the renewal of the fleet. In the late 1960s two brothers equip a 12 m boat, over 20 t, with a very well appointed bridge, a reduced crew, deeper fish holds and greater engine power, a boat adapted to all kinds of fishing; lining, deep-sea trawling or pots on lines. This is the first boat from Houat to fish for scampi, remaining at sea for three days. It is also the first one to go shrimping and catch fish for bait. Then, pot fishermen and shell fishing appear: equipped with more powerful engines, these fishermen work on larger fishing grounds and can practice profitable fishing complementary to dragging for shellfish and scampi. Finally, spider-crab fishing is revived.

This increase in coastal fishing, specialized in pot fishing, is linked to the integration of socio-professional organizations. In order to institutionalize their activity and be acknowledged, the fishermen from Houat form themselves into a craft fishermen group in 1970 and join the union of maritime cooperatives from Morbihan and Loir-Atlantique, the Unicoma. In 1974, the fishermen join the seamen's CFTD trade union, *en masse*.

In the light of these events, the group of Houtais fishermen develop several initiatives: the building of a storage and cooking shop for scampi on the island; the building of a workshop to allow the opening of lobsters; the obtaining of the right to build an earth platform on the port for the supplying of petrol.

From now on, the fishermen of Houat, who are able to rely on a retail price, fixed through common negotiations on markets from Lorient and Quiberon down to Le Croisic, and who can benefit from the equipment and services in ports they had never previously been able to use, make huge investments, taking advantage of the subventions given by the State for the development of small-scale fishing. In the mid-1970s, in a fleet of 28 boats, most of them practising pot-fishing, 89% of the revenues stem form shell fishing, especially crab. Finfish represents no more than 5% of the catch.

This balancing of the pot-fishing is found on a small scale in Mor Bras, a particular region where the inhabitants of Quiberon and Houat are opposed to those of La Turballe. There the fishermen stand by their tradition of sardine fishing, which gave birth to their fishing community, and move *en masse* into pelagic fishing to become specialists in this technique. Introduced in 1970 by the fishermen from Le Croisic in Brittany, it is encouraged by the Fishing Institute and the administration responsible for grants, on account of its 'efficiency': it dramatically raises the productivity of boats and allows a diversification of catches. Due to the volume of fish derived from this fishing technique, the profits are reinvested in larger gear and more powerful vessels.

All the other 'professions' suffer as a result of this type of fishing, the line men especially, because pelagic fishermen catch species like bass and gilthead bream. Retail prices at auction slump. In 1975, the fixing of the minimum retail price (the EU price for sardine and the national price for gilthead bream) favoured pelagic fishermen: the volume of fish captured is so high that, even if the fish is sold at this minimum fixed price, they make substantial profits which they partly reinvest in tourism (camping site, resting places . . .). The gap in revenues dramatically increases between the 'professions', accelerating the process of social differentiation. The tension between people increases; competition becomes acute on fishing grounds, regardless of attempts at regulation by the Administration. From 1978 onwards, the number of serious incidents increases, such as the blockade of ports and the boarding of vessels; the 'pelagic war' is crystallized in the (unintentional) destruction of pots which lie in the way of pelagic fishermen's nets.

In response to this, a new technique modifying the whole deal appears: the 'large mesh net' for deep-sea bottom fish – monk, plaice and turbot. Fishermen form Belle Ile and Quiberon first introduce it in the fishing zone. As a form of pelagic fishing, it is highly selective as it does not allow all sizes of fish to be caught but, conversely, it captures all kinds of species without distinction. The Houat fishermen first set themselves against it, arguing that their technique is really selective because it targets individual species. As they do not mobilize lots of resources they also generate lower returns. How can we stick to this restrictive view, when all we hear about is 'rentability', 'increase in productivity', 'competititiveness', in anticipation of the large European market? And this at a time when economic pressures are increasing (heavier upstream costs and investments together with reduced ranges of prices downstream), while on fishing grounds as well as on the markets there are very different performances and profits.

In 1983, the first islander opens a breach, equipping a 13 m catamaran designed, *inter alia*, to transport the small pelagic trawl as well as the large mesh net. Behind him, six large boats are built to work the same technique: a large mesh net, or possibly a pelagic net, five men on board for periods of two or three days off the coast. Moreover, as the crab, which is used to guarantee regularity of revenue, is hit by a bacterial disease, pot fishermen, whose boats have powerful engines, practice line fishing in turn . . . In the new context of a Common Fisheries Policy (CFP), fishing on Houat will become the victim of its own dynamism.

From 1980 to 1982, the number of boats increases from 33 to 41 and to 45 by 1986. In 1988, it falls back to 41 and thereafter retrogressively to 37 in 1991, 33 in 1992 and 31 in 1993. Several factors combine to explain this decline. The 1987–91 POP (Programme d'Orientation Pluriannuelle; multiannual guidance programme), in the form of the Mellick Plan and its corollary, the PME (authorization for new equipment) play a role at the island level: in 1991 and 1992 only one PME is granted. In 1977, following the bankruptcy and liquidation of Unicoma the group of Houtais fishermen had also lost the help of the co-operative regarding the animation of ideas about coastal fishing, resource management and technical innovation. It was no longer the key element of the island's fishing and it became instead a simple trade organization. Generally speaking, there was no future for young people. The merchant navy facing a cycle of serious crises was unable to provide a refuge.

For everybody, the CFP which applies to all small-scale fishing without any further distinction, in the name of 'resource management'; 'rationalization of the markets' but never in the name of 'a human future', contributes to the demobilization of energy at a critical time: the large mesh net, that has been adopted too late, does not prove profitable. Some fishermen say they will move away from the island. Others hope without really believing, that they will be able to return to seasonal professions and to pot fishing which had originally generated prosperity for the island.

The crisis of the 1990s

Because of the high stakes involved and the chain reactions which are generated, the nature of the latest changes imposed on the Breton coastal communities is very different from previous ones. The current changes are characterized first and foremost by the scale and speed of the transformation involved. No sector is spared – inshore and deep-sea fishing, aquaculture, marketing, etc. A deep shadow is cast over all organizations concerned with maritime activities. All are enjoined to submit to the requirements of new universal rules that are seen to disrupt existing equilibria and cause the collapse of entire areas of practice and custom, which had previously guaranteed survival.

The configuration of change can be summarized in four main tendencies:

- new forms of global organization of production; in particular this includes the 'delocalization' of production which has occurred through the institution of exclusive economic zones (EEZs) and the EU's internal system of redistribution, including quotas;
- the growing share of the market taken by the processing industry, with its own rationale and perspectives on 'delocalization', and the corollary in terms of the transfer of control over added-value from the harvesting to the processing sector;
- the concentration of distribution channels and, in particular, the key role played by the very powerful French supermarkets, where 15 firms sell more than 50% of the fish consumed in France, including 45% of fresh fish, 68% of frozen fish and 88% of smoked fish;
- the progressive narrowing of the links between the processing industry and the retail outlets, thus further constraining the harvesting sector's access to centres of negotiation and decision making in the markets.

Despite recurrent historical references and its use in rhetoric (Delbos, 1989) the issue of resource scarcity is hardly mentioned as a key factor in the crisis.

Here it is necessary to re-emphasize one of the most immediate consequences of these changes, the redefinition of the role and function of the fisherman, who is now an economic actor, integrated into the vertical food system, and cast merely as a commodity producer or supplier of raw materials, without any real powers of negotiation and held firmly in place by the administrative system.

Fish, even including the noble species, are now reduced to the level of raw materials, just as iron ore or grain, allocated by quota allegedly under the rules of a free market. Those same rules also govern the fisherman's life as an economic actor. Supposed to belong to no one and yet to everyone at the same time – even though divided among nations' 'fishing territories' – the resources of the sea have been reduced to a 'common pot' by the actions of the EU. The imposition of quotas and other forms of regulation have rekindled the conflicts between different user groups, competing with each other for the same species – today, for example, the tuna wars with the Spanish over different types of fishing gear; at other times, crab wars with English fishermen . . .

Changing the rules of the game

The changes elaborated above all come within the scope of global processes which alter the traditional relations between man and nature. Because of technical difficulties, it has not proved possible to substitute new scientific 'laws' worked out in laboratories for natural laws – as has been so successfully implemented in modern farming systems (Delbos, 1993). However, other changes have materialized – the delocalization of fishing activity and various technological developments affecting areas downstream from the harvesting activity itself including the transportation of fish. Their effects have been similar, causing a dislocation to the links between the seasons and fishing activities, a key feature of inshore fisheries. For the fishermen the seasons punctuate the natural, biological and meteorological rhythms as the basis for their work routines. Determining the rotations between different activities and the necessary intensity of work, the seasons influence the rates of both natural and human exploitation at the same time. This, in part, explains why fishermen will usually accept 'administrative seasons', which form the dates of opening and closure for different fisheries during the year largely in accordance with nature's seasons, whereas rules governing 'unnatural' quotas and artificial fishing zones are strongly contested and at times ignored.

As it is the market not nature that now fixes the rules of fishing, the fisherman need no longer anticipate the way in which the seasons influence his work. Instead, he should be capable of anticipating the outcome of negotiations between, for example, the former Eastern Bloc countries; Norway and Scotland over the blackfish markets; or between the USA, Canada and France for trade in lobsters, devilfish and shellfish in exchange for cereals or for fishing rights off Newfoundland; or what happens to the cultivation of sea perch in Italy or Greece, or the devaluation of sterling or the Spanish peseta. In other words they may sometimes need to anticipate speculation of currency markets and, at other times, the influence of lobbies on bilateral bargaining.

But such skills of anticipation are precisely the modes of operation of the large processing firms and distribution groups which by these means are able to build up their dominant positions and dispossess the producers of their states as social actors. The processing sector comprises a few very powerful and well organized firms able to achieve low costs for their products through 'just-in-time' management and determination to increase their market share. Thus Unilever makes bilateral deals with former Eastern Bloc countries for supplies, while Findus (or Nestlé) in France will strike bargains with Norway and Iceland.

During the latest crisis, many observers have questioned the capacity of the institutions responsible for the management of the fishing industry to face up to the new challenges. They have speculated on the inefficiency or inadequacy of intervention and regulatory instruments intended to support the industry in adapting to the global changes, as well as the inefficiency of the policies that initiate or encourage those changes.

Internationalization of transportation and currency in a global context of deregulation are most commonly referred to. The mechanisms which regulate world commodity prices appear to be decided without any respect for established rules and are no longer based on criteria relating to the norms of production; they belong instead to the 'no man's land' of the 'market'. Delays in anticipating the outcomes of such mechanisms are explained by reference to the strong growth of production in the late 1980s. But we are also informed of the need for modernization of the entire system that is said to be collapsing under the weight of its 'specificities' and archaic mannerisms, such as the inshore fisherman's wish to continue the traditional logic of 'less fish, higher prices'.

Does not the abstract rhetoric of the market addressed to the 'artisans' of the fishing industry, concerning the globalization of trade and the need for modernization of the industry, also attempt to conceal certain trends? How, for example, public actors like the EU or France simulate the role of private actors – the processing and distribution industries – in the name of public interest. They argue for regularity of supply or for preferential trade with former Eastern Bloc countries, even if it contradicts the interests of their own producers. Private actors meanwhile follow their own logic, depending upon particular circumstances (Proutière-Mouillon, 1993).

Like any institution, the 'market' is a social structure which allows certain groups to assume the power to re-write history on their own terms and persuading others that they can only resign themselves to its influence.

Marginalization and the devaluation of social capital

In the light of these observations, the explosion of anger by the Breton fishing community in 1992 and 1993 appears to expose certain fears, generated perhaps by three particular aspects of the changes experienced by the fishermen:

(1) the narrowing of the socio-economic space in which collective strategies could be deployed – the disappearance of the 'Mauritanian fishery', for example, itself a victim of the same changes – and the consequent confinement of fishing activity to areas which are themselves constrained by a variety of rules; the standardization of norms concerning access rights to the resource;

(2) the growing awareness of the marginalization of the Breton fisherman in his role as economic actor – a process which had remained largely hidden from view by the predisposition to innovation and modernization until the late 1980s, including the success of pelagic trawling in La Turballe in the 1970s (Delbos & Jorion, 1984);

(3) finally, the sudden fall in revenues over the last three years concerning entire production sectors, and combined with severe indebtedness of both enterprises and families caused by borrowing for investment during the previous period of high incomes (Yven, 1991).

The crisis not only exposes the impacts of global restructuring; it has also led to the return of a sense of social precariousness, thought by some to be a thing of the past, and to a loss of symbolic capital – the status and role of fishermen in an urban society – acquired only over the two preceding generations.

A symbolic wound

To face such a crisis, two strategies of social answers have emerged; the first is institutional, under the *aëgis* of organizations and regional authorities, and involves the question of 'fishing crisis' committees in each maritime region in 1993. The aim of these committees, where representation of the professional fisherman is only functional, is to distribute personalized grants to families facing difficulties. The admitted goal of government is 'to calm fishermen down'. In point of fact, these grants have allowed families without income to survive from day to day for four or five months.

The second strategy, strictly endogenous, is based on the reactivation of local associations (the 'families of fishermen', for example) in parallel with the implementation of survival committees. Their aims are to organize solidarity among members; in this respect the associations of fishermen's families achieve significance through the diversity of their actions: the organization of buying groups, common management of money and food donations, sorting out of exchanges of fish and vegetables with local farmers, opening of clothes banks, etc. In this context, the creation of a new area of verbally agreed prices is now beyond national boundaries.

The collective undertaking of its problems by the community is fully assured by all of its members.

On the contrary, the institutional aid on which some families depend is bitterly resented as being incompatible with the values of a society jealous of its symbolic capital. 'We do not ask for charity' was a key expression of the fishermen and their wives whenever they spoke out during the crisis.

The anger of the most exposed categories of fishermen – the inshore skippers and the deep-sea crewmen – reflects their awareness of the marginalization they have suffered in the socio-economic sphere as well as in the institutional world. The absence of any forum in which the redefinition of their roles in the production system might be discussed is tantamount to saying, 'Fishermen, you are no longer social actors; you must make do with the grants you have been offered and with the social measures created to compensate you for the losses you have sustained'.

The response of the fishermen was to create the 'survival committees' and thus forcibly bring the issue of their survival to the notice of the general public as well as to the distribution industry. It is indeed significant that the founders and leaders of these committees also occupy important positions in local and regional fisheries organizations. It is as if they were seeking a new platform from which to voice their appeals for the survival and reproduction of the family-based production systems and to oppose the dominant market forces of institutionalization. But is it simply a matter of reducing the constraints imposed upon them, redefining a minimum level below which the industry cannot survive locally and accepting the reduced role they have been given? Or is it rather the case of a willingness to re-integrate fishermen within the processes of change as fully recognized social actors?

Conclusions

The fishermen's answer, as outlined above, can be interpreted as the struggle of corporate bodies with archaic practices and representatives, whose existence is threatened by the global restructuring of the economy and its consequences, justified only by a particular image of the future. It can also be seen as an attempt by a minority to reclaim their history and thus as a contribution to the development of a new form of citizenship which allows producers to resume their roles in the organization of social life. Is it not the approach of citizen producers to set themselves the goal of retaining control of the circumstances of their economic and social life and thereby to attain the means to influence contemporary changes? Are not the actions of the new organisms, in imposing their social identity and in establishing their place in the organization of human relations (through demonstrations, commando raids and strikes as well as through the dialogue of problems, propositions and solutions) an avant-garde fight? Is the creation of these new representative organizations not an embodiment of the refusal to be confined to the quasi-silent role of 'social utilities', such as those who pretend to be the legitimate spokespersons of the new *Deus ex machina* seeking to rule the world and the market, would wish them to play? Are the fishermen not, in fact, fighting against the 'stabilisation of a social pseudo-natural system that would take place without the citizen's knowledge' (Habernas, 1978)?

References

Boulard, J.C. (1991) L'épopée de la sardine, un siècle d'histoires de pêches. Ouest-France/IFREMER, Rennes/Paris/Brest.

Delbos, G. (1989) De la nature des uns et des autres, à propos du dépeuplement des eaux marines. In *Du Rural à l'Environnement, la question de la Nature Aujourd'hui* (eds N. Mathieu & M. Jollivet), pp. 50–63. L'Harmattan, Paris.

Delbos, G. (1993) L'histoire d'une longue quête, la domestication des poissons plats. In *Cultiver la Mer*, pp. 135–155. Musée Maritime de Saint-Vaast-La-Hougue, Saint-Vaast-La-Hougue.

Delbos, G. & Jorion, P. (1984) *La Transmission des Savoirs*. Maison des Sciences de l'Homme, Paris.

Didou, H. (1994) *Réflexions sur l'Avenir des Pêches Bretonnes*. Rapport pour le Conseil Economique et Social Région Bretagne, April 1994, Rennes.

Habernas, J. (1978) *Raison et Légitimité*. Payot, Paris.

Jorion, P. (1983) *Les Pêcheurs d'Houat, Anthropologie Économique*. Herman, Paris.

Le Bihan, A. (1958) Les préfets du Finistère. In *Bulletin de la Société Archéologique du Finistère*, pp. 3–67. Société Archéologique du Finistère, Quimper, France.

Mollo, P. (1992) *La pêche sur L'île de Houat*. Mémoire de maitrise de l'UF Anthropologie, Ethnologie et Sciences des Religions, Université de Paris VII.

Proutière-Mouillon, G. (1993) *La CEE et le marché des produits de la mer: mécanismes juridiques*. Rapport pour le FIOM et le FROM-Bretagne, October 1993, Paris/Brest.

Vince, A. (1966) *Entre Loire et Vilaine, Étude de Géographie Humaine*. Thése de géographie humaine, Université de Poitiers, Faculté des Lettres.

Yven, A. (1991) *L'Endettement des Familles de Pêcheurs*. Mémoire de Diplôme Supérieur de l'Institut Régional du Travail Social, Rennes.

Chapter 14
Regional Concepts in the Development of the Common Fisheries Policy: the Case of the Atlantic Arc

MARK WISE

Introduction

Conflicts over the development and implementation of the European Union's Common Fisheries Policy are extremely complex, involving a multitude of groups. Intricate regional and sectoral clashes of interest are often simplified into crude inter-state contests with one supposedly homogeneous national group pitted against another; for example, any diversity among 'British' fishermen tends to be obscured when they seek support from compatriots outraged by the image of a new 'Spanish Armada' of Iberian fishermen 'invading' UK waters. These antagonistic national groupings sometimes find common cause in condemnation of politicians and 'Eurocrats' who are often perceived to be imposing, in 'top-down' fashion, an unfair and unworkable fisheries policy devised in remote, centralized European institutions. Such complaints stem from the fact that over the last 25 years or so, the development of the CFP has taken place within the supra-national political structures of the European Community (EC) which, since the so-called Maastricht Treaty of 1991, has become one distinctive component of the wider European Union (EU).

Within this quasi-federal structure, a situation has been created where there is not only the familiar pattern of national conflict over fisheries, but also a constant tension between 'European' and national concepts of fisheries management. The European dimension of a CFP based on law produced by European Union institutions is obvious. It requires that the catching, conservation and trading of fish caught within the 200-mile/median line zones of member states be organized at European Union level. Thus, basic CFP principles require that the free trade of fish within a European single market should be matched by equality of access to EU fishing grounds for member state fishermen free of national discrimination (the so-called 'equal access' provision). Similarly, a common conservation policy has to be developed by EU institutions and applied free of national discrimination by all the member states (Wise, 1984).

However, there have always been dissenters from the view that a common European organization of fisheries provides the most appropriate scale for fishery management. Some argue that fisheries would be better managed at a national level, with each member state being responsible for fishing activities within its own national 200 mile/median line zone. In 1972, and again in 1994, substantial elements in the Norwegian electorate clearly expressed this view when voting to stay out of the EU, not least because of a refusal to accept that Norway's important fishing industry and national fishing zone should be subjected to European jurisdiction. Such sentiments also persist within the EU, notably in Britain and Ireland. This has ensured that national concepts have also shaped many of the CFP's provisions. For example, notwithstanding the 'equal access' principles enshrined in the CFP, six- and twelve-mile national fishing limits still remain wrapped around the coasts of member states with historic rights granted to other countries on a basis which discriminates according to nationality (see Fig. 14.1). Similarly, the total allowable catches (TACs) decided at European level (in conformity with the concept of 'Community Europe') are then divided into national quotas in a manner which is seemingly at odds with the fundamental EU principle that there shall be no discrimination according to nationality (Article 7 of the Rome Treaty).

Partly to relieve some of these 'European-versus-national' stresses, 'regional' concepts have also been introduced into the evolution of EU fishery policy. This chapter aims, first, to examine how a regional dimension has always existed in the CFP and, secondly, to introduce the regional concept of the 'Atlantic Arc' as an example of a new transnational structure which may facilitate a more effective management of European fisheries in the future.

Fig. 14.1   Fishing limits within the European Union.

Regional concepts in the CFP's development

Tension between 'European' and 'national' concepts of fishery policy in the development of the CFP has helped to generate ideas of fishery management which relate to a 'regional' scale other than the Union as a whole or its individual member states. Such regional notions have proved attractive for a number of reasons. First, they provide a way of affording a measure of socio-economic protection to certain 'regional' communities without necessarily breaking the 'European' principle which prohibits national discrimination among member states. Secondly, they offer the possibility of defining spatial entities more appropriate for certain aspects of fisheries management than the Union as a whole or its component national territories. And finally, they respond to political demands, generated by various socio-economic pressures, that have emanated from particular areas in certain member (and applicant) states.

Regional dimensions in the early CFP

Although the ideological underpinnings of the CFP may have been moulded primarily by the biological models of fishery science (Holden, 1994) and 'common-market' economics, socio-economic demands from regions deemed to be especially dependent on fishing have always played a role in the formulation of the policy right from its inception in the mid-1960s (Wise, 1984). Such demands surged to the fore during the first enlargement negotiations (1970–72) between the original 'Six' EC states and the applicant countries of Britain, Ireland, Denmark and Norway. As a result, the original Community's insistence that the *principle* of a fishing access system free of national discrimination be maintained was mitigated by the persistence of the essentially national inshore fishing belts referred to above. However, in conceding these derogations from Community principle, the 1972 Treaty of Accession cast them in regional terms which did not offend too starkly the concept of 'Community Europe'. Thus, the 12-mile limit was extended only to regions adjudged to be particularly dependent on fishing (see Fig. 14.2), while the 6-mile limit was to be reserved, albeit somewhat vaguely, '. . . to vessels which fish traditionally in these waters and which operate from ports in that geographical coastal zone' (Article 100 of the 1972 Treaty of Accession).

Norway's regional proposals in the 1970–72 enlargement negotiations

The Norwegian electorate could not accept the limited and ambiguous exceptions to equality of access included in the Accession Treaty and rejected them in a referendum which prevented Norway from entering the Community. However, during the enlargement negotiations, the Norwegian Government developed the case for a more subtle 'regional' approach to fisheries management lying between more dogmatic 'European' and 'national' concepts. In defence of its fishing communities scattered along the harsh and isolated environment of the country's western littoral, the Norwegian Government insisted that its protective fishery policy formed part of a 'social philosophy' which discriminated not only against foreigners but other Norwegians as well (Wise, 1984). Such discrimination was designed to prevent capital interests based outside the fishing communities from obtaining a dominant position in the fishing industry. If such interests

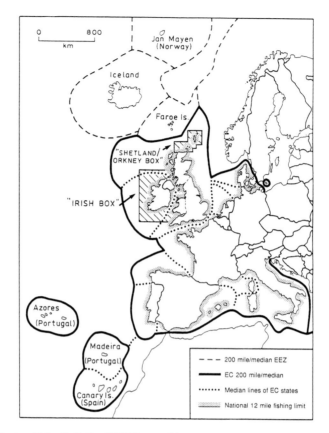

Fig. 14.2   Regional 12-mile limits, 1972 Treaty of Accession.

were to gain a powerful role in Norway's fishing industry, it was felt that fishermen operating on a relatively small scale would be progressively made redundant. This would lead, or so it was feared, to demographic, social and general economic decay over a large part of the country.

In an effort to base this regional approach to fishery policy upon European principles, the Norwegian Government used the 'right of establishment' principle enshrined in the Rome Treaty in order to propose that citizens of other member states could fish within Norwegian waters provided that:

- they were resident in Norway;
- their companies and vessels were registered in the country;
- there was a majority of directors resident in Norway; and
- at least 50% of the capital invested was held by Norwegian residents.

The Norwegians also argued that this 'establishment' proposal provided the most effective means of conserving fisheries. Conservation depends on a sense of responsibility among fishermen for effective enforcement of rules. According to Norway, such virtues are best developed by maintaining a close geographical link between the resource and those who exploit it. Those living in economically precarious coastal communities are more likely, in this view, to protect the resources upon which they directly depend than highly mobile distant-water fishermen who have a less permanent link to any particular fishing ground and are more concerned with short-term gains. Furthermore, while it was thought possible to exercise effective control in Norway's fishing zone from the adjacent mainland where inspectors and inspected lived together, it was difficult, according to the Norwegians, to imagine similar controls being imposed on 'foreign' vessels operating out of distant ports in Britain, Germany or wherever. However, although the Norwegians obtained 'political' assurances in a 'Special Protocol' that its concern to protect the fishing communities of remote western regions would always be taken into account, they were not able to change the basic European 'common market' principles which still lay legally embedded at the base of the CFP despite the derogations attached to them.

Regional concepts in a reformed CFP following the introduction of 200-mile limits

The adoption of 200-mile/median line fishing zones around EC member states on 1 January 1977, creating the so-called 'Europond' (Fig. 14.2.), necessitated reform of the CFP. Over the next six years an entangled political tussle took place, focusing on the basic issue of whether a European Community system of fisheries management should prevail or whether separate national systems should dominate. In general terms, this fundamental conflict divided the relatively 'fish-rich' island states of Britain and Ireland from their less favoured partners. The 'fish-poor' continental countries saw their national interests best served by accepting a 'European' approach which maintained the 'equality of access' principle in a 'Community pond'. This perception was reinforced following the large loss of distant-water opportunities endured by these states during the 1970s. In sharp contrast, the British and Irish insisted on a substantial degree of national preference for their fishermen within the greatly extended national fishing zones around their lengthy shorelines. The differing shares of sea space allocated to states under the new 200-mile/median line regime do much to explain this basic clash between the British Isles and continental Europe, but the UK and Ireland also deployed regional arguments to defend their 'nationalist' stance. In essence, they reiterated the Norwegian rationale about the need to protect fishing communities especially dependent on fishing in their peripheral regions (Wise, 1984).

While rejecting the more nationalistic elements of the British and the Irish position, the Commission was more sensitive to its regional dimension and was thus prepared to temper the Community's equality of access principle with measures to protect local fishing communities and the resources upon which they depended. Thus, in formulating its 1976 proposals for a new CFP, the Commission included among its main objectives a desire to ensure 'as far as possible the level of employment and income in the coastal regions which are economically disadvantaged or largely dependent on fishing activities' (Commission, 1976).

In pursuit of this aim the Commission recognized that '... the straightforward implementation of the principle of equal access is bound to result in the rapid exhaustion of resources ... [and that the] ... consequences of such a situation would be unacceptable' (Commission, 1975). This approach was strengthened by the loss of distant-water fishing rights around countries like Iceland and Norway which threatened a massive diversion of fishing effort into Community waters. This prospect alarmed inshore fishermen who feared competing against large modern vessels for a dwindling resource. Therefore, the Commission proposed that the temporary exception to equal access be extended by creating a 12-mile limit around the whole of the Community coastline for an indefinite period. In acting thus, the Commission stressed how inshore fishing was of 'considerable economic and social importance for numerous coastal regions' estimating that some 80% of the Community's full-time fishermen at that time could be classified as 'inshore' and that the livelihood of some 60 000 people was closely dependent on this sector (Commission, 1976). With this support, the British and Irish were eventually able to win some recognition for the problems of regional fishing communities at a special meeting of EC Foreign Ministers held in the Hague in late 1976.

The 'Hague Preferences' regions

The so-called 'Hague Resolutions' recognized the special problems of Greenland, Ireland and 'northern parts' of the UK where there were communities 'particularly dependent' on fishing. They stipulated that 'account should be taken of their vital needs in applying the CFP'. For example, Ireland obtained agreement that its small relatively underdeveloped fishing industry be allowed to grow according to the Irish Government's Fisheries Development Programme for coastal fisheries which planned to double the national catch between 1975 and 1979. This approach was supported by the aims of the developing common European regional policy. These 'Hague Preferences' were eventually built into the allocation keys announced in the new CFP regulation adopted in January 1983 (Council, 1983a).

Regional 'fishing boxes'

Another concept generated by the tension between conflicting demands for European systems of access and national fishing zones was that of regional fishing plans. The Shetland box (See Fig. 14.2) provides an example of such attempts to organize fishing activities on a geographical scale appropriate to protect specific regional communities and the resources upon which they depend. Within this zone, a degree of preferential access was accorded to local fishermen by discriminating against vessels over 26 metres in length. This non-national discrimination permits the smaller local boats to fish in the box, but larger vessels from further afield must be licensed to operate in this zone. Such an arrangement to protect scarce jobs and resources in a peripheral region seems perfectly compatible with European principles. However, the familiar cocktail of conceptual compromises between European, national and regional approaches can be detected when it is understood that licences to larger vessels are allocated on a discriminatory national basis with 62 to Britain, 52 to France, 12 to Germany and 2 to Belgium.

The large Irish box (Fig. 14.2) created upon the accession of the Iberian states to the EC in 1986 provided another example of a Community measure based on a regional scale. However, its aim of keeping Spanish vessels out of this part of the so-called 'Europond' until at least 1995 in order, among other things, to protect the Irish fishing industry for regional socio-economic reasons clearly discriminated according to nationality, as did the more permanent prohibition on Spain gaining any fishing access to the North Sea arena. Thus, the Spanish have never accepted that this regional box should persist indefinitely and appealed to European principles enshrined in CFP regulations in order to overturn the national discrimination. Eventually, in December 1994, as this regional arrangement came up for review, the Spaniards were able to win access rights for 40 of their vessels to operate in the area, although this was far less than they desired and was obtained in the face of continuing hostility from fishermen in the UK and Ireland.

Regional systems of fishery management and reform of the CFP

The confusion of principles associated with the Shetland and Irish boxes, as well as a more general frustration that the CFP was proving unable to achieve its aim of conserving and allocating fishery resources in a satisfactory manner, has stimulated a debate about what shape a reformed policy should take. Some, while not rejecting a measure of decentralization, argue that European principles be more rigorously enforced in a policy firmly managed by central Community institutions in 'top-down' fashion (Holden, 1994). This finds little favour with member states like Britain which are determined to maintain a strong national hand on fishery policy-making and management. Others suggest a more radical approach whereby substantial management responsibilities are devolved to organizations in the regional fishing areas most directly involved with the industry (Symes, 1994). Already, producers' organizations (POs) throughout the EU are accumulating responsibilities for such things as the local allocation of catch quotas in addition to the market regulation responsibilities they were originally set up to assume. The notion of 'subsidiarity', reinforced in the Treaty on European Union, supports this trend. The Treaty states that:

> '. . . the Community shall take action, in accordance with the principle of subsidiarity, only if and in so far as the objectives of the proposed action cannot be sufficiently achieved by the member states and can therefore, by reason of scale or effects of the proposed action, be better achieved by the Community'

> (Treaty 1992: Article 3b)

The minimalist interpretation of this principle, favoured by British Conservatives, is that it refers only to European Community and national levels of authority. However, others view subsidiarity in a wider sense whereby powers can be devolved to sub-national regional authorities if appropriate. They point to the Treaty's preamble which talks of '. . . creating an ever closer union among the peoples of Europe, in which decisions are taken as closely as possible to the citizen in accordance with the principle of subsidiarity' as well as Article 198 which sets up a 'Committee of the Regions' for the EU. Thus, the scene has been set to encourage a re-examination of what fishery management functions may best be carried out at what level within a reformed CFP to follow the deadline year of 2002. Such a review will doubtless concentrate on regional organizations within member states. However, the possibility of developing transnational regional structures able to facilitate a more satisfactory exploitation of EU fisheries also merits investigation, given the inevitable international nature of much fishing activity.

Transnational functional regions in a future CFP: the 'Atlantic Arc' model

One school of international integration theory, concerned with the resolution of inter-state conflict, urges people to create 'functional' regions designed to deal with specific common problems which cross national boundaries (Mitrany, 1975). Such structures should take a geographical shape appropriate to the task in question. Indeed, functionalist theory envisages a mesh of overlapping transnational authorities of varying spatial extent each dealing with distinctive problems. The history of international fishery policy provides several examples of such an approach. For example, the movement towards the modern 'European' approach to fishery management could be seen as starting when Britain, France, Germany, Denmark, Norway, Sweden, Belgium and the Netherlands saw the practical sense of working together to conserve the dwindling resources of a sea area they commonly exploited, thus creating the North Sea Fisheries Convention of 1882. This led to the creation of other international 'functional' fishery organizations – for example, the North East Atlantic Fisheries Commission in 1959 – geographically shaped to deal with the problem faced (Wise, 1984). Despite the failures of such international bodies to control overfishing, the rationale of trying to find political-geographical structures adapted to deal with the nature of the policy task should not be rejected. Thus, would-be reformers of the CFP could benefit from asking some new questions about possible future spatial frameworks for fishery management. By suggesting the development of transnational regional structures, the 'Atlantic Arc' concept can contribute to this debate.

In simple terms, the idea of an 'arc' extending along the maritime margins of the European Community implies the existence of an area sharing common characteristics, interests and problems stemming from its peripheral maritime location (see Fig. 14.3).

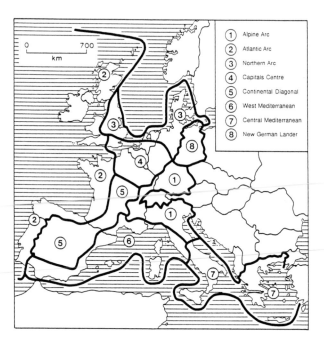

Fig. 14.3   The European Union's macro regions, Europe 2000.

Given this geographical isolation from the main centres of economic growth and political decision-making in Europe, it is argued that the different parts of this maritime belt should co-operate among themselves in order to promote their own regional development by more effectively exploiting and managing their indigenous resources (Morvan, 1989). The overriding aim is to prevent the further economic marginalization of the Atlantic coastal regions in a single European market opening out to countries in northern, eastern and southern Europe. In order to do this, both political and economic co-operation is envisaged along this enormous maritime belt to create a powerful new 'super-regional' entity capable of pursuing the interests of the Atlantic periphery more effectively in competition with the core areas of Europe. On their own, different parts of this maritime belt are, it is argued, doomed to economic and social decline.

This basic idea has already penetrated EU policy-making circles and generated numerous international exchanges among politicians, government officials and business people. A multi-national Atlantic Arc Commission, originating in Nantes but now based in the Poitou-Charentes region of western France, brings together many, but not all, local and regional governments from the whole length of the EU's western facade (see Fig. 14.4). In parallel manner, several chambers of commerce from all the countries of this zone have established their own Atlantic Arc organizations. Out of these organi-

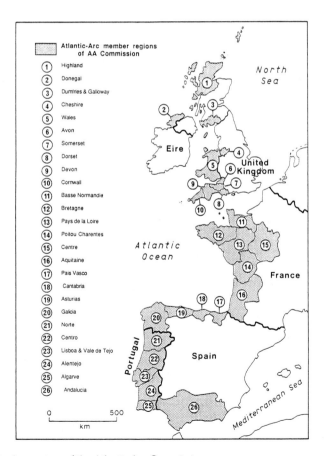

Fig. 14.4   Member regions of the Atlantic Arc Commission.

zations, a multitude of proposals have been made in an effort to convert the aspirations of the Atlantic Arc concept into concrete actions (Atlantic Arc Commission, 1992). Furthermore, in response to such pressures, the European Commission financed a study (CEDRE/ECRD, 1991; Commission, 1994a) of the Atlantic areas as part of its efforts to formulate a future regional development strategy for the Community as a whole (Commission, 1991; see Fig. 14.3).

Inevitably, given its economic, social and political importance, fishing has been identified as one of the many potential sectors where co-ordinated development should develop along the Arc. More than a third of the total EU catch tonnage is made by fishermen from the Atlantic regions, while in terms of value nearly half of the EU landings come from this zone. Furthermore, some 40% of EU fishing capacity is located in the Arc, operated by about 46% of the EU's total number of fishermen. Within the Arc, north east Spain, Scotland and Brittany stand out as major regions of production accounting for about 60% of landings, but also fishing plays an important economic and social role in many other parts of the Atlantic coastal zone (Commission, 1994a). However, whether meaningful measures of fishery co-operation between these fishing regions can develop along this maritime facade remains open to question.

Persistent conflict among fishermen of the Atlantic Arc

Although there is a 'functional' logic suggesting the need for greater transnational regional collaboration among the competing fishermen from different countries who share a dependence on these western waters, conflict is endemic between rival EU producers in this would-be transnational region. Disputes flare up with depressing frequency between national and regional groups throughout the entire length of the Atlantic littoral as fishermen struggle to maintain or increase shares of overfished resources. For example, the 'tuna war' which broke out in the Bay of Biscay during the summer of 1994 was but the latest in a series of such disputes (Eurofish, 1994). Drawing French, Spanish, English and Irish vessels into fierce confrontation, it contained many of the elements typical of preceding conflicts. Spanish tuna fishermen using traditional long lines objected to the continued use of drift nets by other operators, particularly those exceeding the 2.5 km in length stipulated by the EU. Those in favour of drift netting in France, Britain and Ireland argued that much longer nets – up to 8 km – were required to make the fishery profitable. For once, the Spanish, often accused of indiscriminate overfishing throughout the Atlantic Arc and elsewhere, enjoyed support from beyond their shores. These lengthy drift nets not only raised fears of overfishing, but also provoked the ire of conservation bodies such as Greenpeace concerned with the fate of dolphins as well as tuna. The British sought to divert some of this criticism by claiming to use 'dolphin friendly' nets with gaps along their length, but the Spanish were not persuaded by such details and wanted the EU to accept a UN resolution calling for the phasing-out of all drift nets. French fishermen, for whom the Spanish – irked by other disputes in the Bay of Biscay – reserved their greatest hostility, invoked the usual arguments about the social consequences stemming from a loss of income and jobs that would follow a ban on such nets. With the EU's Association of National Fishermen's Organizations (EUROPECHE) completely split on the issue and both European and national authorities unable to take swift preventative action, the conflict degenerated. Rival groups of fishermen intimidated each other at sea; tuna imports into northern

Spanish ports from the drift netting countries were blocked; and fishery protection vessels from Spain, France, Ireland and the UK were despatched to keep the opposing groups of fishermen apart, as well as keeping the Greenpeace vessel *Rainbow Warrior* out of the fray. The situation was made even more complicated by the fact that it was not a conflict fought by homogeneous national communities. For example, the Spanish regional government of Galicia charted its own 'gun boat' to protect its fishermen, while the Basque authorities sought alternative tuna fisheries for their fleet off the Portuguese islands of Madeira and the Azores. Furthermore, fishermen from south west England complained that the UK fishery protection vessels hindered their activities rather than providing protection from the Spanish fleet. On land, meanwhile, the quest for a satisfactory solution meandered inconclusively around European, national and regional institutions as the tuna season drew to a close, leaving the problem ready to erupt again in future years.

Elsewhere in the Atlantic region, similar disputes simmer waiting to be re-ignited. For example, despite the general accord reached in the Council of Ministers in December 1994 about access to the Irish box, the issue of precisely who can fish what, when and where within this extremely sensitive part of the Atlantic Arc has not been fully resolved and rival groups of fishermen remain deeply divided about it. Moreover, such disputes along the EU's western seaboard are not confined to issues of fishing access and fishing gear. International trade relations along the Arc have periodically been disrupted in violent manner. Most notably, fishermen in Brittany and Normandy blocked UK imports which were able to undercut French landing prices as the British currency depreciated in the less-than-perfect EU single market supposedly established by the end of 1992. Furthermore, the direct sales of 'British' fish into Spain by the so-called 'quota hopping' Anglo-Spanish fleet, caused frustration among fishing interests in south west England who saw part of 'their' resources being traded without any benefit to their regional economy.

Quota hopping: an archetypal Atlantic Arc dispute

The clash over 'quota hopping' illustrates many of the interrelated elements found in the fight for access to fishery resources and markets along the EU's Atlantic rim. This phenomenon developed during the 1980s challenging the 'relative stability' concept based on national quotas which was enshrined in the 1983 CFP (Wise, 1984). Broadly speaking, this concept requires that member states obtain approximately the same share of the total EU catch year after year from the zones they have traditionally exploited. This *status quo* would, it was hoped, prevent continuous disputes over fishing rights, conserve stocks and protect communities in marginal regions. The inherent weakness of this allocation system became exposed in the Atlantic regions when Spanish operators bought British boats (which, unfortunately, had not been decommissioned) and registered them in the UK. This enabled them to fish quotas allocated to Britain. Spaniards did the same thing in Ireland as a means of getting access to areas from which they had been ejected following the introduction of 200-mile limits. These boats, although legally British or Irish, were crewed by Spaniards, operated out of Spanish ports, and sold most of their catch directly in Spain without having any meaningful socio-economic link with British or Irish fishing regions. Thus, the spirit, if not the legality, of the 1983 CFP was undermined in that vessels of essentially 'Spanish' character were exploiting British and Irish quotas.

In an attempt to stop quota hopping, the British Government unilaterally implemented a national Merchant Shipping Act (MSA) in the summer of 1988. This required all British fishing vessels to meet much stricter registration criteria clearly based on nationality. Thus, vessels owned by individuals would have to be 100% crewed by British citizens resident in the UK, while vessels owned by companies would have to have 75% of their crews, directors and shareholders resident in the UK. Furthermore, at least 50% of the catch from these UK registered vessels would have to be landed in British ports or transhipped within the UK fishery zone. Finally, they would have to enter a UK port at least four times every three months. Ireland also tightened up its licensing criteria in similar fashion as the fear of Spanish registrations grew. These unilateral national measures angered the Spanish fishing industry, facing serious problems of decline. Consequently, the UK found itself facing a series of court actions pursued by both private companies and the European Commission, under pressure from the Spanish Government. A complex legal battle ensued involving the English High Court and Appeal Court, the British House of Lords and the European Court of Justice.

The British justified their action as a practical measure to establish a genuine socio-economic link between a fishing boat, a fishing state and a fishing quota allocated to that state by the EU. Quotas had been allocated on a national basis in the 1983 agreement in order to ensure that countries and, more specifically, particular regions within them, would get access to the resources they needed without an anarchic 'free-for-all'. The catch reporting problem provided an example of this perceived need for an element of national-regional control in the CFP's management structure. Community regulations required that catches be reported so that the fishing of a particular quota could be stopped when exhausted. However, Spanish vessels registered in Britain, but having no real functional link to a port in that country, landed their catches of British quota in Spain. There was often a long delay before these catches in the Spanish 'landing state' were reported back to the British 'flagstate', if indeed they ever were. When such information belatedly arrived back in Britain, it might be found that the quota had been far exceeded. Overfishing would have occurred and British fishermen in regions like south west England would feel robbed of the national quotas they felt to be rightfully theirs according to the spirit of the 1983 CFP agreement.

The opposing Spanish case was based on the CFP principle that there should be no national discrimination among member states. Thus, the provisions of the MSA restricting registration and crewing rights to UK nationals were in fundamental breach of Community law. Furthermore, in so far as the UK Act prevented Spaniards from setting up companies in another member state, it was also seen to contravene Article 52 of the Rome Treaty relating to freedom of establishment. Other MSA discriminations were thought to be incompatible with the founding Treaty; notably those relating to the freedom to provide services (Article 59) and equality of treatment for shareholders (Article 221). The Spanish also rejected the idea that national quotas were national economic possessions reserved for the exclusive use of particular regional groups. In their view, the CFP required that quota limits be respected in order to conserve stocks, but did not permit nationally restrictive regulations on who could fish them. Indeed, the Spanish maintained that they were often exploiting economic opportunities unwanted by British fishermen in a manner conforming to the common market foundations of the Community. The MSA requirement that vessels dock frequently in Britain and land at least 50% of their catches in the UK was also seen to contradict basic single market aims by hindering the free flow of trade. Ships would be forced to sail to particular ports, thus making them pass their catches through one member state in order to reach another. This denied the economic logic of transporting the product as cheaply as possible to its market.

Eventually, a judgment of the European Court on 25 July 1991 (Judgment, 1991) ruled against both the nationality and residency criteria of the MSA. They were seen to violate the basic rights of Community citizens to exercise an economic activity in any member state free of national discrimination. This verdict was received in sharply contrasting style by rival groups in the Atlantic Arc. The EU Fisheries Commissioner, Mr Marin from Spain, expressed his 'great satisfaction' that the Court's judgment upheld the '. . . fundamental liberties of the Treaty of Rome . . . in the fisheries sector' (Eurofish Report, 1991). In contrast, reaction in Britain was predictably hostile with an MP from south west England, a region much affected by 'quota hopping', deploring a judgment that '. . . makes a nonsense of the CFP, which is based on giving national quotas to individual member states' (Eurofish Report, 1991). Other British politicians worried about the wider issue of UK national sovereignty slipping away towards central EU institutions. A British Act of Parliament had been overruled by the European Court of Justice, thus reducing the UK's ability to manage fisheries in its waters and protect fishing regions.

However, the Court's judgment did not remove all member state powers over fisheries by confirming that:

'It is not contrary to Community law for a member state to stipulate as a condition for the registration of a fishing vessel in its national register that the vessel in question must be managed and its operations directed and controlled from within that member state'.

(Judgment, 1991)

The consequent ambiguity left by this judgment inevitably led to another crisis affecting the 90 or so vessels of the so-called 'Anglo-Spanish' fleet which continued to fish for UK quota in British waters in the face of constant criticism about their alleged inability to respect fishery regulations. As an Anglo-Spanish vessel – the *Blenheim* – was being held in Cornwall pending the payment of a record fine of £331 000 imposed upon it for illegal fishing off the region, a new British effort to contain the quota hoppers broke into the public domain in early 1995. No longer able to require that the skippers of these vessels must be British, the UK Government sought to insist that they speak enough English to ensure that national safety regulations be respected. The Spanish captains concerned were consequently summoned to take English language tests. Predictably this brought familiar protests from Spain that this requirement was contrary to European law forbidding national discrimination. Equally inevitable was the reaction of David Harris, MP for a important fishing constituency in Cornwall. Dismissing the Spanish complaints he insisted that:

'They can't have it both ways. They are claiming they are British boats but they don't want to comply with British laws'.

(*Western Morning News*, 1995)

A potential for fishery co-operation within the Atlantic Arc

Nevertheless, despite the much publicized clashes exemplified by the quota hopping dispute, it is possible to perceive some potential for greater regional co-operation among the competing fishing communities of the Atlantic regions. However, what the geographical scale or scales of this collaboration should be remains unclear. Could the whole Atlantic Arc provide the regional basis for a transnational 'functional' organization within which a greater sense of socio-economic community can be built by getting all competing Atlantic fishermen to co-operate in seeking common solutions to common problems? Could those in the fishing industry of these western fringes respond to the Atlantic Arc ideal and come to perceive a shared interest in developing common strategies to maintain fishing communities in marginal areas where other forms of employment are limited? Pessimists can dismiss such idealism by arguing that national and regional fragmentation of the Atlantic zone is so deep rooted that fishermen are condemned to continue a competitive struggle for individual survival with all the consequences this has for resource depletion and social dislocation. Others, avoiding the extremes of optimism about some ill-defined 'Atlantic community' and pessimism about the destructive side of nationalism, envisage more piecemeal bilateral and multilateral functional arrangements developing among rival fishermen confronted with common problems. But, whatever the nuances of approach, the collaborative possibilities opened up by the 'Atlantic Arc' concept merit further investigation in the effort to find new ways forward for a fishing industry in some disarray.

Indeed, there are already signs that some fishermen in the Atlantic Arc are beginning to sow seeds of co-operation, although it is still too early to determine whether they will bear much fruit. For example, fishermen from south west England and their counterparts from Brittany have reached an unofficial 'gentleman's agreement' about fishing activities in the mid-channel whereby English and Channel Island crabbers (who sell much of their catch in France) are protected from the activities of French (as well as Belgian and British) trawlers. Communication about the resolution of such problems in these shared waters continues in the forum provided by the 'Mid-Channel Potting Conference' which regularly pulls together representatives of the local fishing organizations on opposite sides of the channel. Other serious disputes between English drift netters and Breton trawlers in shared waters have been resolved by regular co-operation between local fishing organizations based in Brittany (Le Guilvenic) and Cornwall (Newlyn).

The EU's 'Pesca' initiative also seeks to develop such co-operative action among fishermen from different countries of the Atlantic regions. It offers grants to facilitate the restructuring of the fishing industry and the development of alternative sources of employment. Among other things, it encourages 'specific projects of a . . . transnational nature in the fisheries sectors' including those designed to promote 'common management of shared fisheries' (Commission, 1994b). Noting the lack of cross-boundary mechanisms to facilitate the discussion of common fisheries management issues, the draft Pesca Programme for south west England envisages a Working Group which would work towards the establishment of a transnational regional organization involving partners from south west England, Brittany, south west Ireland and, hopefully, north west Spain (south west Pesca Programme). This group would seek common solutions to their problems in exploiting shared waters and, in 'bottom-up' fashion, inform the larger debate on fisheries management at both member state and European Union levels. In particular, there is a desire to build transnational regional fishery management structures which could form a fundamental element of a new decentralized CFP system after present arrangements end in 2002.

Similar efforts are being made within the Atlantic Arc Commission to promote co-operative structures arranged by active fishermen in 'bottom-up' fashion rather than imposed 'top-down' from European and national authorities. The aim is not to set up some new Atlantic Arc regional entity to manage all the diverse fisheries from the Canaries to the Shetlands. Instead, the strategy is to use the communication networks developed by the Atlantic Arc Commission in order to facilitate the creation of more local bilateral or multilateral accords dealing with specific problems in a 'functional' manner. Thus,

> '...the existence of the Atlantic Arc Commission can help member regions and their respective sectors of (fisheries) production to get to know each other better and cooperate actively on matters of mutual interest and strive to eliminate conflict...'

> (Atlantic Arc Commission, 1995)

In this context, the Commission has established a 'Fisheries and Fish Farming Working Group' to encourage fishing interests of the Atlantic regions 'to act in unison' by for-mulating proposals between 1995 and 1999 to foster the 'harmonious development' of the resources upon which they depend (Atlantic Arc Commission, 1995). A range of proposals has already been put forward including the organization of inter-regional meetings, first to exchange information and produce joint analyses on fisheries in the Irish box (see Fig. 14.2) and, secondly, to analyse the various fishing practices in the Bay of Biscay fishing season between June and October.

Furthermore, the blockades that have occasionally disrupted the flow of fish trade between the Atlantic regions should not disguise the reality that the Arc takes its most concrete functional form in the substantial commercial links binding largely northern fish producers and southern fish consumers on Europe's western margins. Within the EU's single market, exports from Scotland, Ireland and south west England to the large markets of France and Spain have grown enormously, with the bulk of catches landed in fishing ports like Newlyn in Cornwall destined for more lucrative markets across the English Channel and the Bay of Biscay. Furthermore, direct landings by British vessels in non-UK ports, particularly in north west France, have increased for the same reasons. These trading imperatives also lead to investment in fleets continuing to cross national boundaries in the Arc. The transnational economic forces leading to the creation of the Atlantic Arc's quota hopping Anglo-Spanish fleet persist. For example, in March 1994, the Pescanova company from Vigo in north east Spain took over the 17-strong fleet of *Jégo-Quéré*, France's biggest trawler operator based in Brittany, thus gaining access to 20 000 tonnes of French quota (*Fishing News International*, 1994). A growing opposition among Atlantic Arc fishermen to cheap fish imports from non-EU countries provides another common commercial interest linking disparate groups throughout the Atlantic regions. The growth of these third country imports in response to the demands of fish processors, merchants and consumers is often seen as a threat to the economic and social fabric of Atlantic regional fishing communities which might best be resisted by united action. Thus, a variety of specific fish trade problems provide another potential cement for transnational Atlantic Arc initiatives.

There are already signs that the potential suggested by these economic links within the Arc might be developed. Having already co-operated on other EU development projects, fishing organizations from south west England, Ireland and Brittany are now working together to produce joint commercial proposals within the framework of the Pesca Programme. Among other things, there are transnational plans to rationalize fish marketing chains, promote fish products, seek new markets and co-ordinate fish market withdrawal mechanisms. Discussions on several of these points have already taken place between POs in south west England, Brittany and Normandy (Cornwall County Council, 1994).

Whether these and other initiatives eventually lead to a reduction in tensions between competing groups of national and regional fishermen remains to be seen. Although the Atlantic Arc Commission presents itself as a neutral agency facilitating co-operation, there are those, not least in the British Isles, who view it suspiciously as a vehicle serving French interests. The fact that County Donegal is the only Irish local authority to have joined the Commission indicates that such fears are shared in Ireland (Fig. 14.4). Furthermore, fishermen throughout the Arc are uneasy that the Galician region of north east Spain has been given responsibility for co-ordinating the Commission's Fisheries and Fish Farming Working Group. Despite the tentative steps that are being taken towards more co-operation, mutual suspicion among the Arc's fishermen still runs deep.

Indeed, by the mid-1990s this climate of distrust had led large numbers of British fishermen, not least those in the Atlantic Arc areas, to lend vigorous support to the 'Save Britain's Fish' campaign aimed at withdrawing the UK from the CFP and introducing a national fishery policy based on an exclusive 200-mile/median-line zone. But even if UK fishermen were able to extricate themselves from the CFP – a task of great political, economic and legal difficulty – the need to co-operate with the fishing industry in other parts of the Atlantic Arc would remain. Producers in south west England would still want to sell their fish in French and Iberian markets while fishermen from the more southerly countries would still seek fishing opportunities further north, legally or illegally. More-over, whatever their mutual suspicions, French, Spanish, British and Irish fishermen must find satisfactory ways of organizing their activities in the Bay of Biscay and the Irish box area, otherwise conflicts at sea will continue. Thus, the imperative to create func-tional arrangements to manage access to markets and access to fishing grounds along the Atlantic zone of the EU remains.

Conclusion

This chapter has shown how regional concepts have always played a role in the making of the CFP. Such regional elements have usually been incorporated in an effort to balance the basic European single market aims of the policy with the objectives of protecting fishing communities and the fishery resources upon which they depend. In the debate about how the CFP might be reformed following the deadline year of 2002, the idea that more 'bottom-up' approaches to fishery management involving local regional organizations is enjoying some popularity after the failure of a CFP with a dominantly 'top-down' character to achieve its aims. Those exploring the possibilities of new regional approaches to fishery management should not restrict themselves to consideration of the role that might be played by local producers' organizations and similar institutions within a national framework. The possibility that there may be larger transnational 'functional' fishing regions in Europe, tied together by specific trade links and a common interest in exploiting the resources of particular sea areas, also merits consideration. If such regions – for example, along the 'Atlantic Arc' or around the North Sea arena – can be meaningfully defined in fishery terms, might they not form an element in an overall organizational structure which leads at last to a more rational use of Europe's over-exploited fisheries?

References

Atlantic Arc Commission (1992) Prospective Study for the Atlantic Regions. *AA Chronicle*, **6**, September, Nantes.

Atlantic Arc Commission (1995) *Proposed Business Plan 1995–99*. Cabinet du Président de la Commission Arc Atlantique, Poitiers, France.

CEDRE/ECRD (1991) *Etude Prospective des Régions Atlantiques*. Commission of the EC, Directorate General for Regional Policy, Brussels.

Commission of the EC (1975) *SEC (70) 4503 final*. 22 December, Brussels.

Commission of the EC (1976) *COM (76) 500 final*. 23 September, Brussels.

Commission of the EC (1991) *Europe 2000: Outlook for the Development of the Community's Territory*. COM (91) 452 final, 7 November, Brussels.

Commission of the EC (1994a) *Study of Prospects in the Atlantic Regions*. Official Publications of the EC, Luxembourg.

Commission of the EC (1994b) Notice to the member states laying down guidelines for ... assistance within the framework of a Community initiative concerning the restructuring of the fisheries sector. *Official Journal of the European Commission*, C180/1–5, 1 July.

Cornwall County Council (1994) *South West Pesca Programme: Draft Outline (September 1994)*. Economic Development Office, Truro, Cornwall.

Council of the EC (1983a) Regulation (EEC) 170/83 of 25 January establishing a Community system for the conservation and management of fishery resources. *Official Journal of the European Commission*, **26** (124), 1–13.

Eurofish Report (1991) 1.8.91. Agra Europe Ltd, London.

Eurofish Report (1994) 18.8.94. Agra Europe Ltd, London.

Fishing News International (1994) 'Spain buys top French trawling fleet'. March 1994, p. 60, Heighway Publications.

Holden, M. (1994) *The Common Fisheries Policy*. Blackwell Science Ltd, Oxford.

Judgment of the Court of Justice of the EC (1991) Case 221/89, 25 July, *Official Journal of the European Commission*, C220 23.8.91:5.

Merchant Shipping Act (1988) *UK Statutes in Force*. HMSO, London.

Mitrany, D. (1975) *The Functional Theory of Politics*. Robertson, London.

Morvan, Y. (1989) 'Le renforcement de l'Arc Atlantique: une opportunité pour les acteurs économiques de l'Ouest Européen'. Paper presented at the Assemblée Générale de l'Association Ouest-Atlantique, Brest 16.11.89.

Symes, D. (1994) The European Pond: Who Actually Manages the Fisheries? Paper presented at the International Geographical Union: Commission on Marine Geography Conference on Public Policy and Ocean Development, St Mary's University, Halifax, Nova Scotia, Canada, 12–16 May 1994.

Treaty establishing the European Economic Community (1987) In *Treaties Establish-*

ing the European Communities. Office for Official Publications of the EC, Luxembourg.

Wise, M. (1984) *The Common Fisheries Policy of the European Community.* Methuen, London.

Wise, M. & Gibb, R. (1993) *Single Market to Social Europe.* Longman, London.

Chapter 15
Adapting to the CFP? Globality and New Possibilities for the Faroese Fishing Industry

JÓGVAN MØRKØRE

Introduction

The aim of this chapter is to develop a perspective on the Common Fisheries Policy (CFP) from the somewhat unusual point of view of a semi-autonomous region which forms part of a member state, but which has felt compelled to stay outside the EU because of the Union's fisheries policies. In one sense, it could be argued that the purpose of this decision was to secure 'subsidiarity' in the area of fisheries management.

So far, fisheries politics in the Faroe Islands have not been subordinated to the Union. But things are changing. Because of the industrialization of the fishing industry the islands are today caught in a dilemma in which the debate from the 1970s on whether to join the EC has been revived. At the same time the EU and the CFP have changed, so the Faroese are confronted with quite a different set of options.

The industrial development of the fishing industry has reinforced the processing link in the production process, which in turn has given strength to new pleas for membership or a similar arrangement with the EU.

The Faroe Islands and the European Union: an historical background

In the autumn of 1991 the EC and the Faroe Islands agreed upon a new trade arrangement. The new arrangement replaced one from 1974, which had entered into force after the local parliament (*Løgting*) decided to inform the Danish Government that the Faroe Islands did not want to join the EC. The decision was unanimous. Prior to this historical decision, made in January 1974, Denmark had become a member on 1 January 1973 following a referendum on 2 October 1972.

In the years preceding 1973 when Denmark negotiated with the EC for membership, the negotiations concerned entry for the Kingdom of Denmark as a whole. In the course of negotiations the possibilities for special concessions for the Faroes and Greenland were discussed, but the Faroe Islands could not gain the essential concessions to comply with the strong internal demand for self-determination on issues concerning access to the fishery inside the Faroese fisheries limits.

When, in 1970, the accession negotiations of Denmark, Norway, Britain and Ireland were opened, the original six member states had already agreed upon a fisheries policy based on the main principles of the Treaty of Rome. This included the principle of non-discrimination between EC citizens. In the context of fisheries policy, it meant the same access for all citizens and all vessels to all waters within the fisheries limits of the entire EC.

However, recognizing that such a policy would cause problems in coastal communities along the Atlantic fringe, the Treaty of Accession offered special protection for fishermen from regions such as the Faroes, Shetland, Greenland and western and northern Norway. But these protection measures were geographically restricted to the waters inside the 12-mile limit and, furthermore, there was a time limit of ten years with no warranty at all after that date. As a consequence, no region with fisheries as its major industry deliberately entered the EC.

But, after all, the Faroes were better off than Greenland. Here a referendum was also held with a clear majority against membership. Nonetheless, Greenland was expected to accede alongside Denmark because at that time Greenland did not benefit from 'home rule', unlike the Faroes which had enjoyed the status of an autonomous community within the Kingdom of Denmark since 1948. The Faroe Islands gained some special terms in the Treaty of Accession of 1973. Within three years the Danish Government were to advise the EC whether or not the Faroe Islands were to be covered by Danish membership. The motivation for the three year postponement for the Faroes was said to be uncertainty as to international developments concerning the Law of the Sea, presumably anticipating the widening of sovereign territories beyond the 12-mile limits. But, undoubtedly, the Norwegian decision on membership was just as important a consideration, indeed maybe even more important than the Danish decision itself.

When the biggest competitor on the European markets for fish products decided not to join the EC, the pressure on the Faroes to follow Denmark was relieved. Apparently, Faroese negotiators had hoped for real guarantees concerning the 12-mile limit in the negotiations which followed in 1973, but the EC took a firm stand and refused to make any concessions before the UN Conference on the Law of the Sea had reached a conclusion. At the same time, it became more and more evident that there would be world wide extensions of the economic zones, no matter how the Conference ended. This reinforced resistance to EC membership as the two issues were linked together: EC membership and an extended fishery zone under Faroese jurisdiction became two mutually exclusive options (Neystabø, 1984). The categorical way in which the EC handled the problems in the North Atlantic settled the question. When these were the terms, no political party could recommend membership and the *Løgting* therefore pushed on with the decision not to follow Denmark. Instead it requested that new negotiations on a trade arrangement be opened.

From a political perspective, it could be said that the *Løgting* took the nationalist option to assure Faroese jurisdiction when the anticipated extensions of the fishery zones eventually materialized. There was an alternative option – a unionist one – which put emphasis on the value of the relations to Denmark and fishing rights abroad. But fishing rights in home waters were most important at the time of decision making with the result that a consensus was formed in the *Løgting* based on the nationalist option.

In the resulting trade agreements duty-free access to British markets for Faroese manufactured goods and some fisheries products previously obtained through EFTA was maintained despite the fact that when Denmark left EFTA to become a member of the EC, the Faroes as part of the Kingdom of Denmark was also compelled to withdraw. In Denmark, the traditional freedom from duty for Faroese goods was continued. As for the rest of the EC, an 80% reduction in the duties according to the common tariff was granted for manufactured goods and for certain important fisheries products. With some changes, this arrangement has remained in force until recently.

Since the decision not to enter the EC, the political debate on EC issues has been practically dormant. From time to time there have been warnings from the export company, Faroe Seafood, that things were changing in Europe and that the EC market was becoming more and more important, but it never really succeeded in stimulating a new debate. At least, not until the spring of 1991.

It was pointed out above that Faroese fisheries products were subject to different customs treatment in different EC countries, a fact which was not in keeping with the idea of an 'internal market' (Olafsson, 1989). It might therefore be thought that the Commission would wish to correct this discrepancy.

Taking into account the wider range of fisheries products, a revision of the existing trade arrangements was also in the Faroese interests. Many of these products were not included in the special trade arrangement of 1974. But the politicians seemed reluctant to negotiate a new agreement, fearing that they might emerge with a result that was even worse.

Changing pattern in attitudes

No referendum was held in the Faroes at the time when Denmark decided to join the EC in 1972 and when the *Løgting* decided not to follow in January 1974. Unfortunately, there was no tradition of social surveys in the Faroes until the 1980s, so there is no real evidence as to the strength of the resistance to membership among the Faroese nor to the sentiments towards the EC in the early 1970s. Apparently, it was not considered a political issue at the time. But surveys were carried out throughout the 1980s and 1990s. According to these, the resistance to joining the EU is diminishing, while EU supporters and doubtful voters are increasing as a percentage of the sampled population.

Although a comfortable majority still is against membership, it is no longer an overwhelming majority as it had been in the 1970s and the early 1980s. In a 1984 survey, three out of every four voters were against EC membership, while the latest figures indicate that today less than half the population take a firm stand against membership. Although still modest in proportional terms, those with positive attitudes towards the EU today have increased five fold since 1984 to around 20% of the total.

These trends raise two questions: who exactly are the people changing their stance towards EU and why? Some general answers can be put forward. Employees in the service sector, in particular, have changed their minds over the question of EU membership. The changes at the national level are propelled by some remarkable changes in the central region, Suðurstreym, where the capital Tórshavn is the predominant settlement. Within just a few years, the central region has gone from being the major area of resistance to become the region with the lowest score in respect to negative attitudes towards Europe. Tórshavn – referred to as 'Kontórshavn' ('office harbour') by ironic rural dwellers – forms the major concentration of the service sector, private as well as public. The north east of the Faroes, where the major part of the private capitalist filleting industry and most of the aquaculture is located, shows a similar pro-EU tendency.

The south west has moved in the opposite direction and become increasingly opposed to the EU. In fact the south west has become the new stronghold of resistance to membership. The main distinction between the south west and the rest of the country is the absence of the capitalist filleting industry and the differences in rates of economic growth. Instead, small co-operative plants dispersed among individual coastal settlements have dominated the structure and were heavily reliant on public subsidies. Today, most of these small co-operative plants are closed.

Changing industrial patterns

In order to understand these remarkable shifts, it is necessary to see the changes in a dynamic perspective: economically and politically, as well as culturally or ideologically. Today, it is over two decades since the original EC debate in the Faroes, and within this time major changes have taken place. As a result, the focus of the issues will also have altered. The most significant change has been the industrialization of the fishing industry.

The extension of the exclusive economic zone (EEZ) in 1977 allowed the inshore fishermen to enlarge their catches for a few years. At the time the zone was extended, the cutter fleet was dominant in the Faroese inshore wetfish industry. At that time – in the mid-1970s – their competitors were the British and German trawlers fishing on the banks around the islands.

The world wide extensions of the EEZ to 200 nautical miles pushed the foreign competitors out of the home waters, but at the same time this tendency meant that the Faroese long-distance trawlers, which fished exclusively in foreign waters, were now limited to Faroese waters. So, in a way, the cutters did not rid themselves of competitors. They had to take on new ones. And, although these new competitors were not foreigners, the competition became even more vehement than ever. In the long run, the cutter fleet was reduced in respect of economic importance, number and age. The weaker economic position refers not only to the percentage share of the catch but also to the reduced dependence of the filleting plants on cutter fish (Dali & Mørkøre, 1983). The decline in numbers and increase in age refers to crew members as well as to the vessels (Høgnesen, 1988). One implication of this process is that, today, approximately half the Faroese catch is taken in national waters.

Furthermore, the late 1970s could be characterized as the years of vertical integration within the fishing industry, with mergers between the big plants and the vessel owning companies. At the same time the processing capacity increased as a result of the growth in landings of wet fish. The expansion in processing capacity was due to the establishing of new filleting plants, as well as to investment in new equipment in existing plants (Dali & Mørkøre, 1983).

These developments marked a qualitative shift from a hunting culture to an industrial culture, with a great number of vessels vertically integrated in industrial production processes. As a result, time replaces quantity as the imperative. The production pulse in the processing link imposes new terms on the fishermen. The demand for quality rather than quantity is further enforced by the growing acknowledgement that the resources at hand are limited.

Although this vertical integration deteriorated in a very abrupt way in 1993, it does not mean that the development of quality rather than quantity has been altered. Most of the bankrupted filleting industry was gathered together in one company by the banks and other creditors – a solution which was supported by the Danish Government and more or less imposed on the *Løgting* as a firm condition for further re-financing of the foreign debt. This new company, which has to compete on the world market without being subsidized, renews the pressure for quality as it bids on an open market for fish from vessels which are no longer linked to individual processing plants.

Though fish products still make up 90 to 95% of the exports, as they did in the early 1970s, it is essential to distinguish between exports originating in fish farming and those stemming from catches at sea. In recent years, approximately a quarter of the export value has derived from salmon farming. When EC membership was under debate 20 years ago, fish farming was practically non-existent. But this development does not change the dependence on exports to Europe. In this respect, the dependence on the EU market is more extreme than in the 1970s: at that time less than half the exports went to EC countries, while today some 70% are oriented towards the enlarged EU.

Changing interests – new economic strategies

Comparing the situation in the early 1970s with that of today, the most striking feature was the silence of the principal actors even though there was a lively public debate when the Kingdom of Denmark joined the EC. Today the actors are much more eager to signal their attitudes and articulate their interests and demands in a relaxed political atmosphere. Among economic organizations, the only ones to articulate demands in a manifest way during the earlier debate were the organizations of inshore fishermen which linked membership of the EC together with the question of political and autonomous control of the fishing grounds. At the time their views were articulated mainly through local associations, as the central organization – *Meginfelag Otróðrarmanna* – was not established until 1975.

With their 2000 members, the local association formed a strong pressure group which continuously confronted the political parties with their demands for guarantees for the safeguarding of the fishing grounds. Additionally, a major part of the members of the 'Popular Movement against the EEC' (*Fólkafylkingin móti EEC*) were inshore fishermen (Neystabø, 1984).

The other economic organizations were remarkably cautious, which has forced analysts to focus on latent articulation based on 'objective criteria' such as employment, competitiveness, duties, subsidies, overfishing and so forth (Neystabø, 1984) or to focus on 'national interests' in some sort of a black-box analysis, while abstaining from reflections on domestic disagreements in respect to different economic interests and their impact on the political decision making (Johannesen, 1980). The situation was first of all characterized by these dilemmas. The major organizations within the fishing industry, the Fishermen's Association and the Shipowners' Organization, gave no statements as to their view on the issue. Obviously their members were split because, on the one hand, they were reliant on fishing in foreign areas – including EC waters – and, on the other hand they wanted to gain exclusive control of the Faroese fishing grounds.

The industrialization of the fishing industry implied an integration of the fishing fleet into a vertically structured production system. The economic strategies which accompanied these changes in the 1970s are now challenged, for although the Faroese authorities gained national control over the fish stocks, the sad result has been overfishing.

Since the mid-1970s, particular interests have been highly integrated and mainly engaged in decision-making processes concerning governmental regulatory policies and utilization of fish resources in Faroese waters. Today, fishery management systems are totally reorganized due to overcapitalization of the fishing industry. Apart from mere lobbyism and interest articulation through the political parties, the major sphere of influence for the interest groups in the fishing industry was the formalized corporate system of price fixing and subsidy allocation. This system was suspended in 1990. There seems to be a tacit consensus among all political parties that corporate price fixing and subsidizing have been disfunctional in a very obvious way. The distrust has expelled the interest groups to a body with only consultative power (Mørkøre, 1991a). The way the system worked put an emphasis on the regulatory aspects and the principle that everyone had the same right to catch high priced fish. Consequently, a levelling in prices was necessary to ensure optimal utilization. As the functions were not limited to stabilizing the price, its financial demands on the public purse increased dramatically. Most politicians were defending this tendency with the same arguments as the representatives of the industry, namely that it is the fishing industry which keeps the whole society going. The transfers were not called subsidies but a re-allocation of resources which originally came from the main industry itself.

As a consequence of the absence of market mechanisms, investments in the fishing industry were based on false preconditions leading to growing demands for subsidies to retain an over-capacity and keeping virtually non-profitable sectors alive (Mørkøre, 1991a). Naturally this fishery policy could not go on forever. After the election in 1988 the problems became insurmountable. It became impossible to finance the subsidies either by public loans abroad or by further taxation or savings in the public sector. Hence new economic strategies are being seriously considered in the fishing industry. In these strategies further processing of the fish and competitiveness are the key concepts, while the total collapse of the corporate management regime has made it an open issue as to how the future management scheme should be formed.

Thus the EU issue is placed in quite another context compared to the situation in the 1970s. At that time, the basic prerequisite for the economic strategies was accompanied by a strong resistance to EC membership. But the traditional autonomy within fisheries, which the extension of the fishery zone tried to maintain, became undermined from the other end of the established production system: by the sales companies, which integrated the industrialized production in an international division of labour, where the market is seen to be regulating production in the opposite direction to the production process. As the first link in this chain, fisheries became the dependent part.

The new economic strategies put another perspective on the EU issue. Further processing is dependent on favourable trade arrangements primarily with the Union. Membership could become a necessity. That seems to be the opinion of those within the fishing industry who are closest to the market, namely the sales companies and the filleting plants. By contrast, the fisheries are confronted with the prospect of renewed competition from foreign trawlers on the Faroese fishing grounds, and subordination to EU fisheries politics. The harvesting sector therefore adopts views which are more in accordance with the nationalist strategies of the 1970s.

So far, the dilemma between fishing interests and production interests within the fishing industry has been encapsulated within the vertically integrated concerns. But with the horizontally organized filleting industry without ties to the fishing fleet, the stage is set for clashes between these contradictory interests.

Things are however changing, especially following the economic crisis in 1989 and the subsequent collapse of the management regime. But it was the European Commission which opened the debate in the autumn of 1988 by demanding a new trade arrangement and arguing that the old one would cause trouble when the internal market was established. The Faroese Government, the *Landsstýrið*, responded in August 1989 proposing an agreement on free trade for all goods (Olafsson, 1989). There was no major political debate at this stage. The issue was handled as a purely economic matter and other options, with political implications for self-determination like a customs union and membership of the Community, were not considered at all.

But full support for this policy was not forthcoming. Scepticism was growing, especially among the organizations in the fishing industry. The Faroese Government therefore set up a committee with representatives from each of the major organizations within industry to consider the implications of alternative relationships with the EC.

The committee started its work by circulating a questionnaire to all relevant institutions, organizations and some major companies seeking their views on three options: a free trade agreement, as favoured by the Landsstýrið; a customs union; or membership of the EC. The particular perspective was how to establish these kinds of relationship with the EC as a semi-autonomous society forming part of a member state. This problem is shared by a number of small societies in Europe, as indicated in Fig. 15.1 describing the relations of such societies to the EC and EFTA.

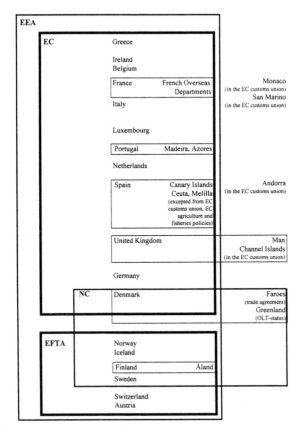

Note: NC = Nordic Council

Fig. 15.1 Semi-autonomous societies with relationships to European co-operation organizations.

The results of the investigation revealed some significant differences between the several component parts of the fishing related economy. Only the union representing unskilled workers was able to lend full support for the preferred option of the *Land-sstýrið*, mainly on account of its opposition to EC membership. Companies engaged in selling fish products were reluctant to endorse such a policy, fearing too many restrictions. During the earlier trade negotiations, the EC had sought to impose import quotas on exactly those products for which the sales companies wanted duty-free entry, in order to facilitate further industrialization and development within the industry. One of the companies, in fact, argued on behalf of applying for membership, should the trade agreement become too restrictive to be properly regarded as a free trade agreement. Most firms within the filleting industry were also open minded about the membership option, although they were placed in a dilemma because of their dependence on a heavily subsidized harvesting sector.

Within the harvesting sector, the vessel owners' organization preferred a customs union, arguing that this would be the best solution for the fishing industry as a whole. They were reluctant to go as far as recommending membership because of the insecurity of exclusive fishing rights in Faroese waters. But in their latest annual report, they seem willing to discuss even membership, relying on the principle of 'relative stability' to provide an appropriate safeguard. Like the vessel owners, the employees' organization also lays stress on the view that a free trade agreement can only provide a temporary solution (EF-arbeiðsbólkurin, 1991).

Thus, it appears that the industrialization of the filleting industry, which resulted from the rejection of EC membership in the 1970s, is today paving the way for a 'yes' vote. The pressure for re-negotiation of the relationship between the Faroes and the EU is increasing largely because of the problems of selling products originating from a third country into the Common Market. Because of overfishing, the filleting plants are having to increase their imports of frozen fish for further processing but are constantly meeting obstacles from the EU. As a result of their reduced dependence on local raw materials, the filleting industry is increasingly distancing itself from the interests of the harvesting sector in putting forward its own arguments for membership.

The conclusions of the committee's report was that the so-called free trade agreement was not an optimal solution. It therefore recommended a new round of negotiations to be held in the near future. In its report the committee advocates either a customs union or a situation in which the Faroes is encompassed by the EEA agreement in a way similar to Iceland (EF-arbeiðsbólkurin, 1991).

Political systems and political strategies

The economic strategies under discussion have political implications and are therefore inevitably connected with political strategies. The political aspects bring new life to the struggle between unionism and separatism, which has dominated Faroese politics this century. But today the scope is wider: 'Union' does not only mean union with Denmark, but with the European Union.

Usually, Faroese politics are characterized by two dichotomies (Wang, 1968; Johannesen, 1980; Neystabø, 1984). On the one hand, the dichotomy is between left and right, or socialism and liberalism. On the other hand, there is the dichotomy between unionism and nationalism, or confederalism and separatism.

This system of differentiation in portraying Faroese politics is often looked upon as unique (see Fig. 15.2). The dichotomy between left and right is, of course, widely known. But the other, the dichotomy between unionism and nationalism, is often perceived as being something especially Faroese. The reason why the Faroese see their politics as being either abnormal or unique is because Faroese politics are being compared to politics in the Nordic countries, but only to those that are sovereign states. This is a mistake, first because the Nordic party systems are especially homogeneous in the sense that they are more firmly linked to class conflicts compared with many other countries in the world, where other cleavages like language, religion, regionalism, or nationality provide party systems with other dichotomies (Lindblad *et al.*, 1984); and, second, because party systems in semi-autonomous areas obviously have to be considered on their own terms (see, for example, Rokkan & Urwin, 1982; 1983).

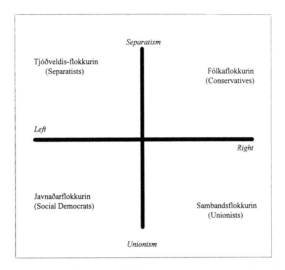

Fig. 15.2   The system of political parties – 1950s to the present day. *Note:* Only the four 'major' parties are taken into account in this diagram. Since the mid-1950s each of them has held a share of around 20–25% of the popular vote. Beside the major parties two, sometimes three, parties are represented in the Løgting. They are normally placed in the area where the two dichotomies intersect.

There is a general consensus that each of the four main parties represented in the Faroese parliament occupies a specific space or cell defined by the intersection of the two dichotomies – left versus right and unionism versus separatism (see Fig. 15.2). The picture presented has remained virtually unchanged since the mid-1950s. From time to time, however, territorial issues provoke nationalist upheavals; this is best illustrated by examining the Separatist Party more closely.

The Separatist Party has always been a populist party. It identifies with 'the people' rather than some particular class; in other words they do not seek to exclude any class. It has, in fact, been closely associated with the fishermen but there is no contradiction between this association and the party's broader 'constituency'. The party's tactics are simply adjusted to suit those classes which express demands in favour of separatism. Those groups in society which form the core of the Separatists' support will thus fluctuate over time. In the mid-1950s, members of the Fishermen's Union formed the most nationalist group, largely through their demands for the extension of territorial waters around the islands. In the early 1970s, it was the fishermen from the small, independently owned cutters who were most strongly opposed to EC membership. The party demanded separation and warned that the constitutional links with Denmark could prove dangerous by threatening to drag the Faroes into the EC along with the Danes. If this were to happen, the EC would exclude the opportunity for Faroese control over an extended sovereign fishing territory and allow British and German trawlers to continue fishing the stocks that the inshore fishermen so clearly wanted for themselves (Mørkøre, 1991b).

Only a few years ago, the Separatists were the only political party to have a clear profile on the EC issue. In the general election of autumn 1990, the party tried to run the campaign on this particular issue. But it failed because the other political parties were reluctant to tell the voters about their future policies on the EC. All other parties simply referred to the free trade agreement which was under way. But the situation was changing rapidly. In the spring of 1991 the Unionist Party decided at their party convention in favour of membership – as an integral part of the Kingdom of Denmark. Thus the polarization between unionism and separatism has been revived in Faroese party politics, although the union now at stake is another and bigger one than simply the Kingdom of Denmark.

From this perspective, the question is whether to give up self-determination in fisheries politics and the exclusive rights to the fish resources around the islands in order to gain access to the important European markets. But it would probably be an oversimplification to leave the question like this. The extra costs arising from tariffs have been estimated at 40–50 million krona a year (EF-arbeiðsbólkurin, 1991). This is a minor sum of money compared to the huge amounts which circulate through Faroese society in the form of subsidy payments.

So surely it would be dubious to argue that it was simply the tariffs which made people change their minds towards the EU. Rather than external conditions, it is the internal structure which is being questioned. The matter at stake is how in the future to organize society itself. Fisheries management, subsidization, giving small communities extra advantages in regional politics – these policies enjoyed a status of consensus since they were based on the self-determination secured in the national compromise of the early 1970s. This consensus is history by now. It is no longer relevant.

So far only the position of the Unionist Party, *Sambandsflokkurin*, is known: it is now in favour of membership. The Republicans will probably have to stick with the free trade agreement, for a customs union would leave the authority to the EU to decide on import tariffs from third countries and an EEA arrangement would probably not allow Faroese political authorities to discriminate against quota hoppers for example. It remains to be seen what positions the other parties will adopt. It is a complex internal matter rather than an external one and the problem to be solved is the economic and political crisis.

In the arguments of the two main protagonists the contours of two competing political strategies are clear. The nationalist strategy argues for autonomy – in fisheries politics at least – as the basic precondition for an economic revival based on an optimal utilization of the fish stocks in Faroese waters. The Unionist strategy argues for EU membership to provide the fish processors with better access to the important EU markets. The nationalist strategy emphasizes the importance of exclusive fishing rights in national waters to deny EU vessels free access to fishing. The Unionist strategy see this as a restraint denying Faroese vessels access to EU waters.

The Nationalists stress the political aspect of democracy and right of self-determination, warning against bureaucracy and the power politics within the EU. This is claimed to be especially important with regard to fisheries politics, since the member states have surrendered the legislative powers over fisheries affairs to the Commission. Denmark, for instance, has very little to say in matters of fisheries regulation compared to the Faroese Home Rule authorities. Put in 'Home Rule' terms, it could be said that Denmark experiences some kind of home rule within the European Union where fisheries have become 'Common Affairs'. By contrast, in the Faroe Islands, fisheries are 'Specific Affairs'.

Likewise in negotiations on fishing rights with third countries, Denmark can no longer exchange fishing rights with other countries. This decision is also reserved for the Commission. Although in principle, Faroese authorities are unable to direct their own foreign affairs, they do have their own policy when it comes to negotiating quotas with sovereign states. Officially, it is the Danish Foreign Office which negotiates the treaties, but in reality it is the Faroese Government that determines their content. In this peculiar way, the Faroe Islands have established treaties with countries that do not have similar arrangements either with the EU or with Denmark: Canada is one specific example.

By contrast, the Unionists emphasize not the potential political disadvantages of membership but the economic advantages that the Faroes may expect to gain from the huge development funds available in the EU. In a sense, we are witnessing a revival of the struggle between nationalism and unionism in Faroese politics. After a decade or so, when left/right issues have been dominant, it seems not unlikely that the other dichotomy will regain its former strength over the next decade or more. So far the two other major parties, the Conservatives and the Social Democrats, have been reluctant to commit themselves on this issue. They face particular difficulties in coming to a view that would reconcile their supporters who tend to be divided on the European issue.

New perspective: globality and second generation agreements with the CFP

The polarization described above leaves politics in a deadlock. Neither of the manifest proposals is able to bring a solution for the fishing industry as a whole nor for the nation, for that matter. Before the restructuring of the industry, when the fleet and the processing plants were still vertically integrated in small concerns, an optimum solution could be found which left no one a loser. The interest groups attached to these sectors were more or less indifferent to the two competing political policies. After the restructuring, this has changed: the filleting plants are now definitely in favour of membership or a system of relationship with the EU that would guarantee similar economic benefits.

A customs union has been under discussion and there are political parties who request that the trade agreement with the EU be renegotiated. They refer to the example of the Isle of Man and the Channel Islands, which lie within the customs union of the EU although they are not members; because of their semi-autonomous status, *vis-à-vis* a member state – the United Kingdom – they have gained access to the internal market. However, it is rather unlikely that a renegotiated agreement for the Faroes would lead to this for the grounds for special arrangements are normally historical rights. The Isle of Man, for instance, was within a customs union with the UK prior to 1973. By contrast, since home rule in 1948 tariff barriers have divided Denmark and the Faroes.

But there are other possible options not so far considered. In the reorientation of the

CFP, the principle of globality is emphasized. This opens up new collaborative structures in order to make sure that so-called external resources will find their way to the European market instead of the UK market, for instance. One way to achieve this is to facilitate joint ventures and support societies in these areas with infrastructure, technology and capital. It is envisaged that a second generation of agreements should combine measures like these with more traditional elements. There are interesting possibilities in this, which could ease down the antagonistic positions between those involved in fishing and those in processing and selling. Efforts have been made to establish the Faroes as a resource centre. These efforts could be made more realistic if the industry were willing to enter into joint ventures with EU capital and if the politicians were willing to rethink economic and political policies in the perspective of these second generation agreements.

New structures: the Barents region as a model for the North Atlantic?

At the same time as this reorientation takes place, new political structures are being designed which might be able to bridge some of the historical gaps. One of the problems for the Faroese is that nation state behaviour results in nation state reactions from the surrounding countries. These patterns are fixed as they have long political traditions behind them. But if the regions are supposed to play a major role in the future Europe, it might be worthwhile investigating and reconsidering the opportunities in such a context. In so doing the Barents region could provide a suitable model for the North Atlantic.

One of the obstacles to co-operation among the fisheries nations of the North Atlantic has been the difference in political level. The problem is how to establish collaborative schemes between a sovereign state like Iceland with semi-autonomous societies like Greenland and the Faroes and again with regions of the sovereign states like western and northern Norway. The Barents region offers an answer to this question, which was raised decades ago but never seriously considered.

The Barents region consists of a two-chamber system. In the Barents Council the ministers of the constituent nation states have their seats, while the regional bodies are represented on the Regional Council. Thus it is possible to create a structure which makes co-operation possible in the North Atlantic. Furthermore the Barents region has been supported by the EU from the very beginning. When the treaty was signed in Kirkenes in January 1993, the EU was represented as an observer and it is still represented as such on the Barents Council. Thus the Commission has committed itself to this project. Likewise, it should be possible to integrate fisheries politics and regional politics in the North Atlantic to the benefit of both the EU and the independent fisheries nations which so far have been forced to stay clear of the Union. This may possibly provide an appropriate context for a revised CFP; it will certainly require a rethinking in the north.

References

Dali, T. & Mørkøre, J. (1983) *Stat og Småborgerligt Fiskeri i den Faerøske Samfundsformation.* Institute of History/Institute of Political Studies, University of Copenhagen.

EF-arbeiðsbólkurin (1991) Føroyar og EF – Útlit Fyri Samvinnu. Føroya Landsstýrið, Tórshavn.

Høgnesen, O.W. (1988) Samfelagslig lýsing av útróðrarstœttini v.m. Lecture at 'Útróðrarstevnan í Klaksvík', Conference for the Inshore Fishermen, Klaksvík.

Johannesen, K. (1980) *Faerøsk Fiskeri – og Markedspolitik i 70'erne.* Institute of Political Studies, University of Århus.

Lindblad, I. *et al.* (1984) *Politik i Norden: En Jämfórande Översikt.* Liber Förlag, Stockholm.

Mørkøre, J. (1991a) Et kroporativt forvaltningsregimes sammenbrud – erfaringer fra det faerøske fiskeri i nationalt farvand. In *Fiskerireguleringer*, pp. 44–82. Nordiske Seminar og Arbejdsrapporter, Copenhagen.

Mørkøre, J. (1991b) Class interests and nationalism in Faroese politics. *North Atlantic Studies*, **1**, 57–67.

Neystabø, K. (1984) *Faerøerne og EF.* Egið forlag, Tórshavn.

Olafsson, A. (1987) Uttanríkisviðurskifti Føroya/Faerøernes undenrigsrelationer/The foreign relations of the Faroe Islands. In *Fiskivinnuábókin '87/Fiskeriårbogen '87/ Fisheries Yearbook '87*, pp. 33–36. S.J. Repro, Tórshavn.

Olafsson, A. (1989) Føroyar og EF/Faerøerne og EF/The Faroe Islands and the EEC. In *Fiskivinnuárbókin '89/Fiskeriårbogen '89*, pp. 53–63. S.J. Repro, Tórshavn.

Rokkan, S. & Urwin, D.W. (1982) *The Politics of Territorial Identity: Studies in European Regionalism.* Sage, London.

Rokkan, S. & Urwin, D.W. (1983) *Economy, Territory, Identity: Politics of West European Peripheries.* Sage, London.

Wang, Z. (1968) Faerøsk politik i nyere tid. In *Faeringer-Fraender: Sprog, Historie, Politik og Økonomi*, (ed. A. Ølgaard), pp. 76–110. Gyldendal, Copenhagen.

Wang, Z. (1988) *Stjórnmálafrøði.* Futura, Hoyvík.

Part V
Fisheries in a Global Food System

Chapter 16
The European Fishing Industry: Deregulation and the Market

PETER FRIIS

Introduction

There is no doubt that the sources of supply for the European fish markets are becoming increasingly internationalized both in terms of the processing industries and the retail markets. The effect of this development is of considerable significance to the traditional primary producers – the domestic European fishing industries – which no longer occupy the role of sole suppliers. The fishermen are losing power in this process and, increasingly, they are no longer the principal actors. As a result, the focus must be shifted from the fishermen and the processing industries to other access of activity – in particular, to the service industries and primarily within the distribution sector.

Since the volume of the world catch has fallen, relative to the size of the consumer market, the expectation is for higher first hand sales prices. Such simple predictions, arising from the relationship between supply and demand have not in fact materialized. Internationally, prices for fish have dropped and the European markets no longer enjoy protection. This internationalization of supplies has a number of negative consequences for the primary fishing industry. The question is whether the fishing industry is in a position to affect this development. The chapter explores a number of basic propositions relating to the problems facing the European markets and the domestic fish industries in greater detail; it focuses particular attention on the impact of trade deregulation and the abandonment of protection for the agricultural sector and on the consequences of the supermarkets' increasing dominance of the European food retailing industry.

In the current situation, the European fishing industry is unable to supply the European market with the amount of fish demanded

The reason for the shortfall in supplies is partly due to the increasing scarcity of resources in the north east Atlantic, particularly for whitefish which form the basic primary produce of the European fishing industry, and partly to the growth in consumption. The level of self-sufficiency, in terms of fish and fish products, has fallen drastically over recent years – from 83% in 1984 to only 57% in 1992 (see Fig. 16.1). As for the most important primary produce of the fishing industry, blocks of frozen fillets, the fall in the level of self-sufficiency in Europe has been even more dramatic; today, it stands at under 10% (see Fig. 16.2). The reason is that EU fishermen are

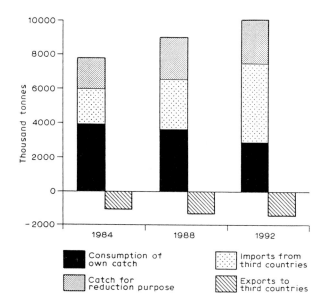

Fig. 16.1   Supply balance of fish and fish products, European Union, 1984–1992.

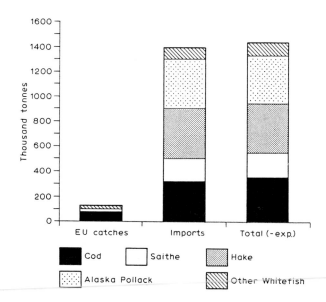

Fig. 16.2   Supplies of whitefish for frozen blocks, European Union, 1992.

unwilling to sell their catches for this purpose at the very low prices which are on offer.

The evidence suggests that for a number of years the cost of catching fish has been higher than the market was willing to pay

The costs of catching fish have been rising rapidly in recent years and, according to recent estimates (FAO, 1992) these now exceed the gross landed value of the catch by a staggering 44% (Fig. 16.1). One reason for this huge imbalance between gross market revenue and costs is the fact that many countries now subsidize the costs of fishing: the former Soviet Union, the Faroe Islands, Norway, Greenland and the EU have in the recent past paid out huge sums in subsidies. Fig. 16.3 indicates the very steep rise in the level of support by the EU for its fishing industries in the late 1980s.

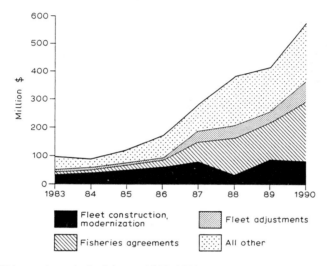

Fig. 16.3　EU financial support for fisheries, 1983–1990.

The price mechanism has functioned as an Achilles' heel in negotiations between the industry and the mechanisms of regulation

The price of fish is a crucial factor in determining the success of enterprises engaged in the harvesting sector and will influence the future size and structure of the industry. The effect of falling prices is equivalent to a reduction in earnings and the low prices, combined with restrictions on the total allowable catch, imply that there is room for fewer fishermen within the industry. In the long run, therefore, current price levels will drive down operating margins and many more fishermen will be forced out of business.

International deregulation will lead to a further fall in food prices, including those for fish and fish products

Following a briefly sustained peak in 1988–90, prices for fish and part-processed fish products have been falling, despite the fact that the total world catch of whitefish has fallen (see Fig. 16.4). Recent international trade negotiations, including GATT and the European Economic Area (EEA) agreement, between the EU and the EFTA countries, together with the reform of the EU's Common Agricultural Policy, have contributed to the trend towards lower commodity prices. On the international level, there seems to be a broad consensus that in order to escape the general trend towards global recession, nation states must continue to reinforce their efforts for further deregulation of international trade.

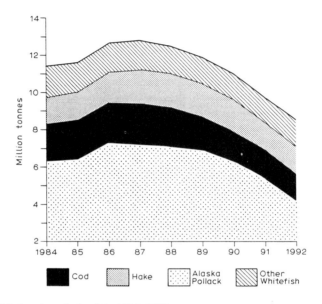

Fig. 16.4   Global catches of whitefish, 1984–1992.

The recently concluded GATT negotiations discussed issues that had earlier had been considered political dynamite, namely the international trade in food products. One of the issues concerned fish and, although the precise outcome is not yet quite clear, one thing is certain: current levels of import duties will be reduced – many by as much as 50% – over the five-year period starting in 1995. The EEA agreement with EFTA countries has also resulted in significant reductions in import duties levied on fish imported into the EU, with a few significant exceptions. The purpose of this deregulation is to reduce the EU's price subsidies for food products, so that prices may approximate to those on the world market.

The background to agricultural reform in the EU

The need for reform of the Common Agricultural Policy lay partly in the fact that the EU's agricultural support schemes had got badly out of hand and partly because the latest round of the GATT negotiations required the Union to put forward initiatives for the reduction in the level of agricultural subsidies. Since the 1960s, the prior goal of the CAP had been to produce as large a quantity of food products as was possible, initially in order to make the EU self-sufficient in food production. The means by which the goal was to be achieved was a combination of guaranteed minimum prices for agricultural produce, set independently of world market prices, and a system of import duties and export subsidies for agricultural commodities.

The goal of self-sufficiency was achieved in the 1970s but the system of protection remained in place. In subsequent years the mountains of surplus production, affecting grain, butter and meat in particular, grew. The major problem of the agricultural support scheme was that the subsidy was allocated according to the volume of production: the more the farmer produced, the larger the subsidy he received. In 1992, the EU countries produced 120% of their food consumption needs; agricultural subsidy schemes took up 60% of the EU budget. For several years, the EU has attempted to reduce surplus production by restructuring support towards non-food production, including voluntary set aside agreements, early retirement schemes and so on.

One of the major problems faced during the most recent Uruguay round of the GATT negotiations was the issue of trade in agricultural products. In this context particularly, the EU's protective policy was seen to obstruct the path to agreement, as none of the remaining negotiating parties were willing to accept the Union's agricultural subsidy schemes nor the tariff walls preventing access to the European markets for agricultural products.

Reform of the Common Agricultural Policy

The EU's new agricultural reforms were agreed on 21 May 1992. They were to become effective from 1 July 1993 and be fully implemented over a three-year period up to 1 July 1996. The fact that the reforms were agreed before the GATT negotiations were concluded may imply further changes to the reform package. The most important features of the reforms are:

- a switch in subsidy support from production (tonnes produced) to area sown, regardless of the quantity produced;
- a drastic reduction in grain prices of 29% over the three-year period, coupled with the substitution of compulsory set aside for the earlier voluntary scheme;
- a further reduction in milk quotas of 2% and in butter prices of 5%; and
- a reduced ceiling for intervention storage for beef from 750 000 t in 1993 to 350 000 t in 1997, with a concurrent reduction in open market prices by 15% over three years.

The package clearly signals a considerable and very rapid reduction in revenues for the European agricultural sector. But there will be other significant knock-on effects to food prices resulting from these changes to agricultural support; one very important change – with potentially serious repercussions for the fisheries sector – is the 29% cut in grain prices which will greatly reduce production costs for pork and poultry.

Competition for the market place

Patterns of food consumption are normally highly sensitive to price variations. When buying food products, most consumers will weigh quality against price. It is therefore difficult to imagine that sales of fish will increase or even hold their market share unless fish prices manage to keep pace with those of alternative food products which compete directly with fish. In principle, this applies to a whole range of food products competing with whitefish, but in particular to the lean 'white meats' like pork and poultry which have recently commanded the consumers' attention.

Globally, the consumption of poultry meat has increased dramatically at the expense of the traditional red meats. Consumption is particularly high in the USA, though it is rising most rapidly in Japan (see Fig. 16.5). Today, annual European consumption of chicken stands at around 20 kg per person and is increasing, though it still falls far short of levels in the United States and significant variations in national consumption levels are clearly evident (see Fig. 16.6). Americans consume over twice as much as the average European and sales of convenience cuts continue to prove very successful, as has the consumption of chicken in terms of fast foods. In recent years, sales of turkeys – and especially fresh turkey – have increased dramatically in parts of Europe (see Fig. 16.7). The overall success is due partly to falling prices, but effective advertising campaigns promoting poultry consumption have also had a considerable impact. By contrast, promotional advertising for fish consumption is rare and, when it does occur, it is nearly always associated with the promotion of particular brand products created by the large multinational food manufacturers.

It is important to emphasize that the consumption of fish cannot be viewed independently of consumption and price movements in general. It is expected that the prices

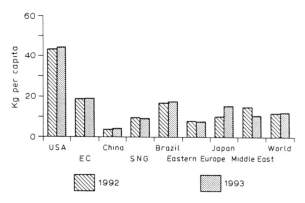

Fig. 16.5 Per capita consumption of poultry, selected areas, 1992 and 1993.

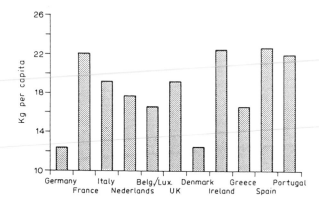

Fig. 16.6 Per capita consumption of poultry, European Union, 1991.

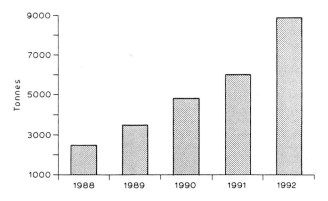

Fig. 16.7   Consumption of turkey meat, Denmark, 1988–1992.

for a large range of food products which compete with fish will fall over the next few years. At the same time, the price of whitefish is unlikely to increase significantly, but whether consumption levels can be sustained by parallel price reductions and/or improved generic advertising remains very doubtful.

Price formation for food products is increasingly a consequence of the growth in the negotiation powers of international retail chains: buyer-seller relations are becoming increasingly asymmetrical

From once being a national matter, price formation has now become an international concern; from formerly being determined by producers and resources, it is increasingly being decided by consumers and retailers. Among the most important changes influencing the state of the markets for fish are those that characterize the relations between buyers and sellers. A key determinant of the changing balance of power between the producers and the retail trade is the nature and location of knowledge concerning the market (Burt, 1990)

Until the mid-1970s, the food producers, by and large, were in a position to determine the type of commodities and products marketed by the food retailers. Detailed knowledge of consumer preferences had been relatively unimportant in the immediate post-war market situation, characterized by a shortage of goals and by relatively few mass-produced goods which sold rapidly. Gradually, however, producers sought to extend their decisive monopoly of knowledge about consumer behaviour and preferences: they paid market research firms and established consumer panels in order to enhance this knowledge, which was exploited for product development and to target their growing advertising budgets. But market analysis proved very costly and was limited to only a fairly small segment of consumers; it thus tended to provide only fragmentary knowledge of consumers' demands and desires. In many cases, market research surveys were only commissioned in connection with the launching of new products (Anon., 1994; Friis, 1993)

Today, the balance of power between producers and retailers is greatly altered, with increasing concentration of the retail trade in the hands of large, powerful, multinational chains (see Fig. 16.8). These retail chains have exploited their close relations with the

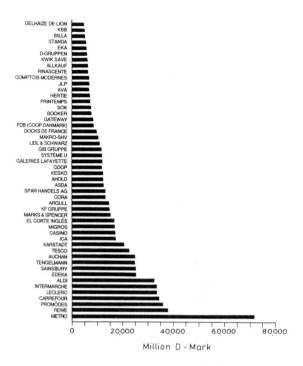

Fig. 16.8 Europe's 50 largest retail chains, by turnover, 1992.

consumers in order to obtain an advantage over the food producers in terms of a much more detailed and comprehensive knowledge of consumers' quality demands. The retail chains are, moreover, in a very much stronger position to exploit the benefits of information technology in terms of accounting systems, stock control, staff management, etc.

The introduction of information technology, in the form of bar coding and electronic points of sale, means that not only are the supermarket chains in a position to calculate the precise turnover rates for specific articles and groups of articles; they can also calculate the optimal amount of shelf-space for each individual supermarket and, without any extra cost, they can develop complete profiles of customer behaviour month by month, week by week and day by day. It has thus become possible to 'experiment' with customers' reactions to any changes in packaging, brands, quality and price. The retail chains' knowledge of consumers' behavioural patterns and preferences in relation to both traditional and novel goods has become a decisive parameter of their advantage. No individual producer is in a position to obtain similar knowledge and, without such knowledge, the producer is left to engage in product development, as if it were wearing a blindfold. It is the redistribution of knowledge that has largely altered the balance of power in the negotiations between producer and buyer.

On the other hand, most retail chains are inexperienced in production or product development. Various forms of collaboration exist between retail organizations and the producers, particularly the food manufacturers. Previously, 'collaboration' often took the form of invitations to bid for sales contracts, with the lowest bidder winning the contract. Today, more advanced forms of collaboration of an intimate nature, implying considerable mutual trust, may be characterized by mutual exchanges of information. But the prime initiators and innovators are the retail chains.

The exploitation of knowledge

In the late 1970s, the retail trade discovered a major means of challenging the producers with the introduction of their own brands (distributor brands, own labels, private labels). The recession, meanwhile, forced consumers to be more attentive to price (Dawson *et al.*, 1986) and their loyalty to producer brands weakened. The retail trade has increasingly pursued this strategy as it can satisfy several objectives. It meant, for example, the availability of goods that were cheaper than the traditional producer brands – cheaper because the retail chains did not have to pay advertising costs. Often, they were in a position to achieve larger profit margins than the producers of brand name articles. In addition, the retail chains could use their own brands to profile themselves in competition with other retail outlets and so gain customer loyalty.

There were considerable differences in the way the various retail chains marketed their own brands and in the prestige that attached to them. For some, like Marks and Spencer, the Swiss Migros and the Danish Irma, the retail stores will carry only their own labels, while others will sell producer brands alongside their own labels.

Consumers are no longer as loyal to brand name articles as was previously the case; they are much more prone to 'zap' between brand name articles, the retail chains' own labels and discount products. The last two are still increasing at the expense of brand name articles. The profit on brand name articles is often very small or even marginal. Nonetheless, many retail chains regard them as an indispensable ingredient of the concept of selection.

For their part, brand name manufacturers will react differently to the retail chains' disinclination to carry their articles. They are obliged to focus their efforts on retaining or possibly increasing consumer loyalty in order that their products be represented on the supermarket shelves among the retail chains' own brands. Their strategies have included:

- marketing campaigns in the media;
- consumer coupons;
- gift coupons (Nestlé/Walt Disney figures);
- the development of low-fat products for the discount segment (Findus);
- redirection of some of the production into the retail chains' own labels;
- price reductions; or
- a combination of several of those strategies.

It is difficult to predict the outcome of the war between retail chains and the brand name manufacturers in the different European markets. There are unfortunately no reliable statistics to illustrate the trends, but the retail chains' own labels seem to gain ground particularly in those countries where they already hold a strong position. This applies especially to food products. The British marketing consultancy, Euromonitor, has shown that from 1988 to 1992 sales by private levels increased by more than 11%, while producers' brands grew only by 6.5%. There are, however, some major regional variations: the retail chains' own labels have not, for example, gained a position in Scandinavia in any way similar to that in the UK, Germany and Switzerland where they are already well established. Euromonitor's analysis of the war between manufacturers' and retailers' labels does clearly demonstrate that the retail trades' own labels are gradually ousting all the famous brand names. Organizations like Marks and Spencer and Migras, whose share of own labels is high compared to their total selection of goods, will serve as an example for other retail businesses.

However, within a surprisingly wide range of product areas, top brand names have managed to increase their market share; this applies to perfumery, soap, hair care preparations, tinned soups, cat food and french fries, *inter alia*. The retail chains are attempting to expand their own label in image-creating products such as frozen foods and body care products as a counter measure to the segment's gradual decline.

In general, since the early 1990s, the change in market power favouring the large corporations or retail chains, together with the intense price wars, has gradually undermined the primary producers' very chances of business within the food industry. A recent study, published in *Lebensmittal-zeitung*, on the 'duel of the brand names' suggests that the share of own labels within food production will have doubled by the year 2000. But not all food manufacturers are suffering from this trend: large conglomerates like Nestlé, Unilever, Kraft and Philip Morris are clearly still able to survive.

Recently, it has become clear that the retail trade in Europe has assumed the role of interpreting the consumers' demands upon the food industry. Manufacturers, small and medium sized, have been subject to strong pressures, which are now increasingly applied to the large producers of brand names. As a result, a relatively small number of retail chains and their purchasing organizations will continue to act as trend setters for developments within the food industry. Power and control in relation to marketing and product development will increasingly lie within the realm of the retail trade as the essential link between the food industry and the consumers. Furthermore, the structural development, characterized by still fewer and larger internationally oriented retail corporations, implies that the retail trade will continue to make more and more demands on their suppliers in terms of prices, quality, selection, marketing and the use of technology.

The general global conditions and development trends are expected to exert still further influence over the suppliers of food products. The worldwide stagnation in consumption narrows the marketing opportunities. Internationalization of the basic resources of primary production, such as agriculture and fisheries, increases cross-border competition and promotes an international division of labour. Finally, the internationalization of the retail trade adds to the retail sector's negotiating power and thus contributes to the internationalization of consumption patterns. In consequence, the primary producers continue to lose influence within the vertically integrated food chain.

It is the retail trade that informs the producers what it is that the consumers want and, due to the retail chains' increasing purchasing power in terms of their own labels and the international collaboration in purchasing supplies, the producers are increasingly forced to adapt their production to the demands of the retail trade.

The days when buyers and sellers of fish were both regarded as 'experts' are gone. Purchasing organizations are increasingly populated by economists or persons with commercial training. To them, accredited guarantees of quality are often more important than the precious relations built on trust between buyer and seller. As a result, the purchasing organizations prefer certified producers – those, for example, with ISO 9002 standards to their name. A number of Nordic manufacturers of fishery products, including Norwegian peeled shrimp processors and Icelandic fish exporters, have already been drawn into the consequences of this development, while in Denmark it has tended to be the large processors of herring and the owners of herring boats that have obtained the relevant certification.

The retail chains' growing role as fish sellers

In those countries like the UK and France, where food retailing is already dominated by supermarket chains, their role as fish retailers is becoming very important. In the UK, the multiples already account for more than 90% of total sales in canned fish, 74% of frozen fish and around 37% of wet and smoked fish (Gentles, 1993). In France, the multiples' share of total sales in frozen fish is 63% and for smoked fish their share is as much as 88%. The turnover in fresh fish has also risen substantially in the last few years, with the growing role of the hypermarkets – large supermarkets of more than 2500 square metres of retail space – with extensive fresh fish departments. In France, the multiples distributed 47% of all fresh finfish in 1993 (Montfort, 1993).

Rivalry among retail chains – in fish sales as well as in other food products – can be counteracted through the creation of strategic alliances and purchasing collaboration. Studies show, however, that in the next few years we are likely to witness further dramatic concentrations in food marketing, partly because the total turnover in the food sector is not expected to increase significantly (Jørgensen, 1989; Montfort, 1993).

Air freight poses a new threat to the European fishing industry

The core activity of Europe's domestic fishing industries is the supply of high quality fresh fish to the retail markets. But this area too is now threatened by the development of regular air freighting of fresh fish from around the world. Fish is airlifted to the European markets from three countries: hake from Namibia and Argentina and, in January 1994, Icelandair commenced regular deliveries of fresh redfish fillets through Luxembourg bound for the German retail market. Across the Atlantic, the amount of fresh fish entering the US market by air from New Zealand and Australia is also growing rapidly. Although the volume of air-freighted fresh fish reaching the European markets at present represents only a very small fraction of the total market sales, the new trend underlines the vulnerability of the European fishing industry stemming initially from the very low first-hand sales prices for good quality fresh fish.

Conclusion: the organizations' response

The European Union's fishermen's association, Europeche, has recently put forward its own solutions to the current problem of low prices for fish: the deregulation of fish imports from third countries and the introduction of higher minimum prices for domestic landings. Unfortunately, the proposal contains two particular problems. In the first place, it fails to take into consideration the fact that the general price level for food products is falling and will probably continue to fall by approximately 20% over the next few years, following deregulation of the agricultural sector; it is difficult to imagine that fish prices can remain immune from the reduction in prices for food products in general.

Secondly, international experience suggests that subsidizing food production through high guaranteed prices leads to over production and, in the case of fisheries, to over-fishing.

According to the evidence presented above, it is clear that two specific factors have exerted a decisive influence on price formation in the European fish markets over the past few years: first, deregulation of international trade and the removal of protection from the agricultural production system; and, second, the very powerful negotiating position in the food markets held by the retail chains and their international purchasing organizations. It is unlikely that either of those situations will be reversed.

To restore something of the former status of the domestic fishing industries in Europe, one may need to abandon the image of fishing as an industry that heaps up fish on the quayside whenever circumstances allow. Instead, it could be transformed into an industry capable of supplying the markets with high quality fish especially during periods of less plentiful exogenous supply. Adjusting the fishing industry to the markets in this way would certainly necessitate an improvement in information systems linking the boats to the markets and so making the connection between supply and price formation more apparent.

References

Anon. (1994) *Forbrugsgoder, – en Erhversøkonomisk Analyse.* Erhvervs-fremmestyrelsen, København.

Burt, S. (1990) *The Internationalization of European Food Retailers.* Institute for Retail Studies, Glasgow.

Dawson, J., Shaw, S., Burt, S. & Rana, J. (1986) *Structural Change and Public Policy in the European Food Industry.* FAST Occasional Papers, No. 103–4., Brussels.

FAO (1992) Marine Fisheries and the Law of the Sea: A Decade of Change. *Fisheries Circular*, No. 853, FAO, Rome.

Friis, P. (1993) EFs marked og dets betydning for de nordiske fiskerisamfund. In *Fiskerisamfund-Hvilke Veje?* (ed. G. Gudmundsson), pp. 15–38. NORD 1993:27, København.

Gentles, P. (1993) The place of seafood and retail development in the UK. In *New Markets for Seafood* (eds W.F.A. Horner & G.A. Rodriguez), pp. 1–25. University of Hull International Fisheries Institute, Hull.

Jørgensen N. (1989) *Jagten på den Forsvundne Fordel. EFs Indre Marked og Dansk Landbrugs og Foraedlingsindustris Mulighed.* Handelshøjskole Syd. Kolding.

Montfort, M.C. (1993) The French market for seafood – recent developments and prospects. In *New Markets for Seafood* (eds W.F.A Horner & G.A. Rodriguez), pp. 26–34. Unversity of Hull International Fisheries Institute, Hull.

Panorama (1994) *Yearbook from the EU about the State of Industries in the Community.* Brussels.

Chapter 17
The Geopolitics of Fish: the Case of the North Atlantic

ÖRN D. JÓNSSON

Introduction

The chapter deals with the future of the North Atlantic area from a geopolitical view-point whereby specific attention is paid to fish as a product and to the changing producer/buyer relations. The underlying rationale of this approach is due to the ambivalent traits of the North Atlantic as a regional concept in economic, political and cultural terms. The discussion starts by examining the changing socio-economic relations of the region. The geopolitics of fish will be analysed as well as the emerging world market for fish and fish products. The specific characteristics of fish as a product will be emphasized and the discussion is taken to its conclusion with three alternative aspects of the possible future: crises, catastrophes and possibilities.

Interpretations of the North Atlantic region

A region or a geographic area is, by definition, an integrated part of a larger whole. It can be argued that the development of a given geographic area and its actual situation at any point in time is just as much the result of overall world development as the result of local initiative. This statement becomes increasingly true with increasing globalization.

The geographic location of the North Atlantic is fairly clear, though the practical side of mapping its topographic location can of course be disputed the nearer one approaches the North American continent. That is to say, the overlapping of the geographical definition of the area does not converge with the socio-political division between America and Europe. In that sense my definition is basically pragmatic, based on the fact that I have ashamedly little knowledge of the areas that belong to Canada. For me, in all practical contexts, the North Atlantic is Greenland, the Faroes, Iceland and (northern) Norway. In a word, it is European rather than American. It is much more difficult to define the socio-economic situation of the area. The interdependency with other parts of the world has changed dramatically over the last decades in connection with overall changes of the political and economic world order.

To start with, it is important to keep in mind the phenomenological fact that every community is, in a sense, the centre of the world. At the same time we have to take into account that most communities have been forced into the world economy rather than having consciously decided to integrate themselves in the larger whole. This is parti-

cularly true for latecomers to the modernization process. And, in addition, the latecomers in most parts of the world have tended to be more vulnerable to the modern
industrialized world's thirst for perpetual change.

Historically, it can be said that the North Atlantic countries have been a meaningful
part of the Nordic countries. To a certain extent, this assertion refers to the fact that
Greenland, the Faroes and Iceland were former colonies of Denmark or still are. Norway
is not only Nordic, it is Scandinavian. In the post-war period, when the USA emerged as
the superpower of the world, both the political role and buying power of Europe was
temporarily reduced. At that time the North Atlantic was 'comfortably' located in
between Europe and North America, both in a political and economic sense.

With the emergence of less developed countries, the economic upsurge in the Far
East and, recently, the disintegration of the USSR and Eastern Europe, the socio-
economic world map has been redrawn and consequently the North Atlantic has been
'relocated' on several levels.

The USA's relative decline meant the re-emerging importance of the Nordic alliance.
For the countries in the North Atlantic, including Greenland, Iceland and northern
Norway in terms of fish products, Denmark has played a central role as a gateway to
Europe and even to the rest of the world. This is partly due to Denmark's membership of
the EU, but is also based on a long tradition of a division of labour between the countries
involved. In a quite complicated way, ties with the Far East, mainly Japan, have been
strengthened. To support this statement it is sufficient to say that Greenland and Iceland
are probably among the few countries in the world that have a positive trade balance
with Japan.

The critical question today must be the relationship between the North Atlantic and
the EU. At a certain point of time in the not so distant past, it seemed that Iceland and
Liechtenstein were the only Western European countries not applying for membership
of the European Union. At that point of time, Greenland and the Faroes were rather
imprecisely situated inside and outside the Union. Today the picture is somewhat
clearer. Finland and Sweden have elected to join the Union, though Norway has decided
to remain outside. The complex and somewhat critical acceptance of the Maastricht
Treaty has meant that life outside the Union is possibly a viable alternative.

Nevertheless it is clear that the concentration of political and economic power in the
central regions of Europe is an historical fact that will continue to exert a profound
influence on the development of the North Atlantic area. This is partly reflected in a
recent study commissioned by the Union which came up with the theory which commonly goes under the name of 'blue bananas', a tainted map of Europe where two
banana-like structures were drawn around the most developed regions of central Europe
(Ceccini, 1988). In a recent study by the Nordic Regional Research Institute an attempt
is made to develop a Scandinavian variant of this theory by drawing several bananas on
a map of Scandinavia, principally the Baltic and the Barents regions (Veggeland, 1992).
This endeavour was probably intended to diminish the peripheral consequences of the
blue banana concept, but at the same time functions as an acceptance of this moder-
nized version of the centre-periphery theory.

Bananas or not, it would be ridiculous to try to draw similar structures on a map of the
North Atlantic region. Greenland, Iceland, the Faroes and, for that matter northern
Norway as an entity, do not fulfil the criteria of either an economic or a political whole
required to become a strong union to the same extent as the most developed regions of
continental Europe – in other words, to justify the banana tag. In the future we will

probably see more cooperation in the area, but that should rather be looked upon as an addition to the present international relations rather than a substitute or alternative. It is hard to visualize the new colour of a Norwegian/Faroes ice green banana!

To sum up, what we are experiencing is the increasing importance of Europe in a dangerously centre/peripheral way, whereby the North Atlantic is losing its central position on the socio-economic world map due to the diminishing impact of the superpowers. At the same time, increasing ties to the Far East – due to mutual interests in the fishing business – are quite significant.

The geopolitics of fish

Fisheries and fish processing are becoming an increasingly global endeavour, where developments in some areas of the world have a considerable effect on others, in spite of the vast geographical distances.

Over-exploitation of fishing grounds is a nearly universal fact. The remedy to the 'tragedy of the commons', the 200-mile fisheries zone, has had a more profound and long-lasting effect than people envisaged during the fight for territorial rights.

Distant-water fishing is rapidly decreasing. Germany has scrapped its factory fleet to a large extent. Spain tries to hang on but rather to achieve a better negotiating position in the European Union on regional matters and fishing rights than to revitalize its distant-water fishing. The fishing fleet of the former Soviet Union is in disarray, and Japanese distant-water fishing has decreased by half in the past three years (Ministry, 1993). The other side of the coin is the emergence of new fishing nations on the world scene: nations that have rich fishing grounds which now are being exploited by national capital or on a joint-venture basis. Chile, Argentina, Peru, Kamchatka and Namibia are just few examples. Significantly, Iceland, Norway, and – to a limited extent – the Faroes have each been involved in the modernization of the fishing industry in at least some of these countries.

The decline of distant-water fishing has meant an increase in the world trade of fish and fish products. This trend has been further strengthened by the growing importance of aquaculture.

The paradox of the growing importance of the fishing industry on the world market level is that, within the various nation states, fisheries tend to be looked upon as secondary or even peripheral in the national economic context. In the more affluent societies of the world, fisheries are often treated as parentheses to the general food production system or to agriculture. And recently fisheries have tended to be defined as a part of regional revitalization strategies in the EU, a structural instrument intended to diminish the painful consequences of the centralization of political and economic power. This is also true in the case of Japan, Taiwan and to some extent the USA and Canada.

From a geopolitical view, or in a socio-economic context, what unites the North Atlantic countries is their overall dependency on fisheries as the main source of export income: a situation that makes them somewhat unique in the complicated array of world fisheries and processing. Historically this has meant that the countries of the North Atlantic have been competing with each other with similar products on the same markets.

The emergence of a world market for fish products, or rather fish as a raw material for processing, may change the interrelationship of countries in the North Atlantic area.

One aspect of the globalization trend is the weakening of ties between catching and processing. The vertical integration of firms from the fisheries to the market, which has served some of the North Atlantic countries well, especially Iceland, is now becoming a thing of the past. The nations in the area have, to an extent, changed roles from being solely sellers to become both buyers and sellers. This is of course most obvious in the case of northern Norway where the close geographical proximity of the Russian border has completely changed their socio-economic position in less than two years. A more absurd example of this world market in the making is the import of orange roughy from New Zealand into Iceland for processing before being sold on the French market.

The emerging world market, the loosening of the ties between catching and processing, and the emergence of new able players on the world market, means fiercer competition as well as falling prices for standard fish products. This trend, further accelerated by changes on the buying side, will be discussed below.

Fish as a product

With a fair degree of simplification, it can be said that the fisheries as a business can be defined by the ambivalent term of industrial hunting. The North Atlantic has been strong in the supply of raw materials and semi-fabricates for those nations that are in an under supply situation in their home markets for fish. In this area the North Atlantic is losing ground. And, with the important exception of Norway, the North Atlantic has not been able actively to participate in the ever more important industry of aquaculture. On the positive side, the North Atlantic countries have been successful in gaining shares of the, hitherto, burgeoning market of high quality frozen fish, namely ground fish and shrimps frozen at sea.

In order to understand the North Atlantic nations as suppliers of fish products, it is important to look into the specific characteristics of fish as a product. In contrast to many other food products, fish is usually most valuable in its natural state of unprocessed freshness. Here it is important to underline the second characteristic mentioned: freshness does not only refer to the fish as 'wet fish', i.e. not frozen, dried or salted. The time from the catch to the time of consumption or processing plays a crucial role in the determination of the quality of the fish.

Fish as a product is one of the few food products that have increased in price in the post-war period. This fact has been explained by many converging factors such as the changing priorities of consumers to a more healthy consumption pattern, increased affluence of the nations that have cultural preferences for fish products, overfishing in some of the most important fishing grounds and the extension of fishing limits to 200 miles as a nearly universal system in the world of fishing.

In the last two or three years there has been a significant tendency towards lower fish prices. This tendency is quite visible in Europe and at certain points has led to severe conflicts between fishermen, officials and importers of fish (see for example, Chapter 13). At first the increasing flow from the Barents Sea region was blamed for the instability of the European fish markets, but as things became more complicated the people involved have realized that the problem is of a more complex international nature.

Part of the explanation lies in the increasing segmentation of the world fish market or, rather, the polarization of fish products as, on the one hand, standard industrialized commodities, and 'natural' luxury objects, on the other. We see a strong trend towards

'Macdonaldization' of society and, at the same time, a 'back to nature' trend preferred by the middle and upper middle classes or those who can afford to live a 'natural' life in a polluted and ever more crowded world. The polarization is all the more interesting in the light of recent market analyses that concluded that the luxury consumer and the cost conscious rational consumer can be one and the same person. People tend to submit to fast food in the rush of the working day; on weekends or special occasions, the banquet is more exquisite (see Fiddes, 1991; Ritzer, 1993).

The general trend towards lowering of fish prices is partly to be found in the development of these two somewhat separated segments. More able players are entering the market for industrialized fish. A leading figure in the Danish fishing industry commented that chicken was probably the ultimate fish product. With the increasing importance of supermarket or hypermarket chains, the distribution channels of fish are moving from the fishmongers to the cooler or the freezing cabinets. The preferences of consumers in most countries is leaning towards convenience and the characteristics of the most convenient fish products are boneless fillets, preferably mild in taste, white and pan-sized. In the catering business cod and sole have been the products that fulfil these criteria. But under the present conditions any mild tasting whitefish will do. Pollack from Alaska, which is much more abundant than cod or sole, and whiting from South America seem to pass as the 'chicken of the sea' under certain circumstances. This trend is well known in the USA and Canada where Alaskan pollack has been gaining ground for several years. The luxury or gourmet market has been severely affected by the economic downturn in most countries. The 'yuppie' is becoming cost conscious even in his moments of conspicuous consumption.

In the long run, a more profound change will influence the situation of the North Atlantic as a supplier of fish. The image of fish as a product is changing along with the conception of food products. From the Middle Ages until quite recently the European preferences in food could be characterized as 'red, white and green'. Red meat was absolutely the most preferred food, while white meat, fish and poultry came in second; grains and greens had a considerably lower status. Now the development is towards 'green, white and red'. Where vegetables, fruits and grains have been elevated to the premier level; white meat holds the middle, while red meat is becoming suspect on criteria of healthy heating. But again, this is a trend limited by factors related to class, ethnic preferences and age. In magnitude, the industrialized fish products are winning ground in the more affluent societies, and this product is taking on the characteristics of a generic product, where origins and species are becoming less important than the basic qualities of homogeneity, regularity of supply and convenience. In this category, which accounts for the bulk of the fish products consumed, fish (without regard to its origins) and poultry stand on equal ground. The competition is primarily related to price and, to some extent, quality.

The luxury fish is an image product; but the 'chickenfish' is an industrial product where the following factors seem to be of crucial importance:

- globalization means more able players;
- improvements in transportation and information technology imply changes to the relationship;
- hypermarkets throughout the world tend to press prices close to production costs; and
- brand loyalty is losing ground to 'own labels'.

For the North Atlantic these development trends mean that the region has lost its role as the absolutely dominant supplier of 'fresh' and frozen groundfish. The question then is if, and to what extent, the region can become a supplier of fish and fish products in the luxury category. Is the concept of wild catches from unpolluted waters of the High North a realistic possibility or just wishful thinking?

Crises, catastrophes and possibilities

The present situation can be categorized as follows:

- the North Atlantic fisheries community is becoming an increasingly integrated part of the world;
- the area has lost its role as the dominant supplier of certain 'raw materials' for the affluent nations;
- the future is determined by three factors – ecological conditions, supply and segmentation, or the interplay between crises, catastrophes and hindsight.

The crises from an economic point of view are due to the general downturn in the global economy. Consumer preferences have changed and the guiding light is value for money. This has meant that fish, although it is regarded as a healthy alternative to red meat, is more expensive than vegetables, fruits, grains and poultry which are the other commodities that have healthy characteristics attributed to them. The increase of able suppliers of standard industrial fish products has resulted in an overall fall in prices and fiercer competition in the higher margin segments. The positive intonation of fish as a healthy product will probably mean that it will regain some of its status if the world economy starts to recover.

What is catastrophic about the situation in the North Atlantic region is the permanent overfishing and the question mark many scientists raise over how renewable the resources really are. In recent discussions the combined problem of overfishing and ecological changes has been put into focus. It was one of the main themes at ICES (1993) symposium on *Cod and Climate Change* held in Reykjavik. Climatic changes can possibly have a more significant effect on recruitment of such central species as cod than we were originally led to believe. Diminishing catches mean less room for manoeuvre with regard to production of fish products. The future of the region is primarily determined by the possibility of implementing functioning fisheries management systems. No fish, no fish products, no affluence.

But, as always, it is necessary to look for the light at the end of the tunnel. In my view it is a rather unfruitful endeavour to present instant solutions for an area as vast and as politically, economically and culturally different as the North Atlantic region. There are many and diverse possibilities of course. But with regard to fish and fish products the concept of freshness plays a central role.

- The big supermarket chains and the catering industry have been stressing a policy where they do business with suppliers as close to the source as possible.
- Although fish frozen at sea means diminishing work on shore, nevertheless its success is based on the superiority of the product. The prices of products frozen at sea have dropped considerably less than most other fish products.
- Advances in transportation and information technology have and will continue to

have a considerable influence on the development of the fishing industry. Among other things, this will change the buyer-seller relationship where networks based on trust will be more frequent.

- As fish products become a more integrated part of the general context of modern consumption behaviour, through the new distribution channels of the supermarkets, the door opens for product development. This means that fish products are becoming more knowledge intensive; and knowledge in this context means, with a little twist of irony, the art of selling less fish for a higher price where the consumer is buying packaged image in the name of the superior service. Producers in the North Atlantic have shown that they can participate in this game in limited areas of the food market.
- Globalization is both a threat and, at the same time, an opportunity, opening up several possibilities for the area. The fisheries are amongst the most advanced in the world; the processing is quite advanced and the educational level is high. The North Atlantic could become an active player on the world market in fishing, processing and even in marketing of fish products. Also the production of technological equipment for the fishing industry has proven itself to have some potential.

Conclusion

The North Atlantic is becoming an integrated part of the global fishing community, which means:

- if your product is not distinctive in some way or another, take the given price;
- the skills and technological knowledge in some parts of the area are becoming a product in itself;
- the North Atlantic has to follow the fish to the world market;
- fish products and fish are not synonyms.

The future is therefore based on three questions:

- To what extent is it possible to keep the renewable resources of the sea on an acceptable level?
- Can the North Atlantic countries find other means of keeping their standards of living at relatively high levels than simply their closeness to rich fishing grounds?
- What will the relationship to the EU be? Or, how does Brussels define the colour of a North Atlantic ice green banana?

References

Ceccini, P., Catinat, M. & Jacquemin, A. (1988) *Europa '92, Realiseringen af det Indre Markéd*. Europakommisionen, Copenhagen.

Fiddes, N. (1991) *Meat, a Natural Symbol*. Routledge, London.

Hersoug, B. (ed.) (1992) Fiskerinœringens Hovedtrekk – Landsanalyser av Danmark, Fœrøerne, Grønland, Island og Norge. Nord 1992:30, Copenhagen.

Ministry of Fisheries (1993) *Annual Report of the Ministry of Fisheries*. Ministry of Fisheries, Tokyo.

Ritzer, G. (1993) *The Macdonaldization of Society*. Pine Forge Press, London.
Veggeland, N. (1992) *Regionalpolitikkken Udfordres*. NordREFO, Copenhagen.

Conclusion

Chapter 18
Sailing into Calmer Waters?

KEVIN CREAN and DAVID SYMES

All the contributions to this volume, from a variety of viewpoints, describe and analyse the condition of crisis that currently affects the fisheries of the North Atlantic and Mediterranean. The authors highlight the decline in the quality of life and standard of living of many of the littoral fishing communities and the sense of frustration experienced by fishermen in the face of the contraction of fishing opportunities. The message comes across that it is a crisis borne out of the failure of fisheries management strategies founded on biological and economic goals and that this policy approach has resulted in the needs of the fishermen being overlooked by government and the resource regulators. In many situations fishermen have responded independently and, in the eyes of the state, illegally, to safeguard their livelihoods. The regulators have reacted by imposing more restrictive regulatory systems that have not necessarily been more effective but, together with global economic forces, have quickened the pace of change within the fishing industry. Indeed, an apt quotation from Chapter 3 explains that depth of institutional change required to cope with the implementation of management strategies based on bio-economic objectives has led to:

> '... processes where people are actually deprived of their capacity to act in a constructive and 'rational' way. They are left with strategies where stability is restored through the formation of groups based on aggression towards outsiders and destructive action.'

Inevitably the net effect of these actions has been to exacerbate the failure of existing policies aimed at the management of fisheries. The question that now needs to be addressed is whether we can look beyond the foreshortened horizon of crisis and despair for any signs that the issues surrounding fisheries management both regionally and globally are becoming any less turbulent.

Beyond reiteration of the crisis statement the contributors have sought to describe and analyse various facets of the fisheries management dilemma within the common theme of managing change in a turbulent economic and political environment. In line with the failure of policy, attention has been drawn to the poor performance of some existing institutions, particularly in relation to the track record of *traditional* institutions in the EU and elsewhere in the fisheries world.

Within the main theme it is possible to distinguish several sub-themes embraced by the discussion. First, there is the plea, on behalf of fishermen and their communities, that the strategies for fisheries management should take account of social objectives. The task of undertaking this change must not be underestimated as it will require a searching review of policy making at a fundamental level; whilst countries such as Norway have developed a consultation system that feeds social aspects into the policy-making mechanism, this is unusual in terms of other European nations. Indeed there is the concern that institutions involved in the development of management strategy for fisheries resources are not equipped with the planning methodologies and resources to effect the changes desired (Crean, 1994). However the Mo-i-Rana case study provides a workable model that could have potential application for the promotion of change in the fishing industry. Although inevitably the chances of success would hinge on the involvement of the players in a strategy in which social issues are a priority, legitimacy is created and orderly and understandable industrial reshaping takes place.

Second, in line with the need to ensure the involvement of fishermen in policy making, there is a need to develop robust institutions that have an *internal* 'architecture' (Kay, 1993) that is sufficiently flexible to cope with change. Furthermore it is a widely held view that the *external* 'architecture' of the fishermen's institutions should articulate positively with the institutions of government in general and resource regulation in particular. Devolved management is popularly viewed as a device to promote a work-able external 'architecture' between government institutions and those of the fishing industry (Jentoft, 1989).

Third, there is evidence that the appropriation of local management roles by the state threatens civil order and the breakdown of those social relations which hitherto have helped to mediate processes of modernization within the fishing community. The debilitating effect on local fishermen's organizations clearly robs both the fishing industry and the local community of an important positive dynamism. At least one of the con-tributors is somewhat skeptical of local or regional fisheries management *per se* but does argue that regional alliances, both political and commercial, can re-establish a platform for the articulation of the fishermen's needs in a more relevant context. The majority of the contributors to this text emphasize the need to come to terms with the problematic issues that surround the sub-theme of property rights. In the North Atlantic fisheries, the inequitable and discriminatory aspects associated with the deployment of quotas and ITQs and also restricted access areas are highlighted as priorities for the early attention of the fisheries institutions.

Finally, our attention is drawn to the macro-economic forces that drive the global business environment and consequently pose a threat to functioning of the fisheries institutions. Here some of our contributors have discussed possible changes that might lessen the unfavourable impact of what more often than not is a turbulent business environment.

Prior to drawing together ideas on the formulation of a research agenda for the future the editors have some final remarks on three key areas that are identified as influencing the development of fisheries policy in the future – co-management, property/user rights and fisheries on the high seas.

Co-management

In the aftermath of the crisis discourse that characterizes many of the contributions, authors have wrestled with the conundrum of what steps to take to improve the situa-tion? In Chapter 3 Pálsson & Helgason give a clear direction in their statement addressing the way forward in response to the crisis:

'We may well be advised to search for alternative epistemologies and management schemes, democratizing and decentralizing the policy making process.'

Indeed this message, although phrased rather differently, comes through from several of the contributors who perceive that within the sub-theme of institutional development, the concept of co-management is seen as a mechanism to manoeuvre the European fishing industry out of crisis. Others have taken the argument further and enquired whether it would be possible for the EU to intervene in the current problems of the fishing industry taking a more positive role and that this should entail the:

'relocalization of management decisions concerning the design of institutions to the level of the coastal community or fishery-dependent region'.

If such a devolution process were to take place then the question should be posed as to what is intended by the process of co-management? Some contributors felt that fishermen and their dependent coastal communities were looking for an organizational structure in which policy instruments are deployed that might optimize income, employment, influence and resource 'ownership'. However to make progress in this direction there is no doubt that a heavy load will fall on the institutions that seek both to manage fisheries and represent fishermen (and their dependent communities). In Chapter 4 Sandberg warns of the cost of failure:

'if there is poor correspondence in the distribution of duties and rights, the resource management institutions will crumble from within because of a lack of legitimacy. Here fishers will often "take back their rights", and "black fishing" and "black trading in fish" will flourish and the cost of government control will mount'.

Further weight to this prediction maybe found in the accounts of the struggle for survival of, respectively, the traditional fishermen's organizations (*cofradías*) in Spain and *prud'homies* in France with, the relatively new producers' organizations (POs). Several contributors are at pains to stress the 'legitimacy' of the traditional organizations and by inference question the validity of the 'modern' organizations. Clearly, in the process of institutional restructuring the aspect of 'legitimacy' needs careful consideration, and in terms of the devolved management of fisheries resources a number of characteristics come to the fore that newly emerging organizations might do well to emulate if they are to gain acceptance. Traditional institutions are notable in that they have:

- a geographical (often contiguous) link with the fishing communities and fishing zones;
- a functional structure aimed at defending the social interests of the whole fishing community;
- a legal basis in the law of the State;
- the capacity to apply technical measures as part of the fisheries management process;
- the ability to allot resources equitably amongst their members.

A further endorsement of the importance of the legitimacy issue comes from a study of the Fisheries Co-operative Associations (FCAs) in the management of fisheries. The coastal waters of Japan have for centuries been under co-management regimes with fishermen participating in the formulation of fisheries policies, allocating fishing licences and playing active roles in the implementation and enforcement of regulations. In a number of ways the FCAs compare with the some of the traditional institutions of EU – the *cofradías* and *prud'homies* – although arguably the FCA represents one of the most successful examples of devolved management in the world fisheries. Yet there are quite profound differences that are once again potentially instructive in the development of devolved management strategies. Could this be because success of the FCAs in the devolved management of coastal fisheries is a function of the 'continuity of the institutions' and that the 'customary regulations (of the littoral fishing communities) have been incorporated into formal law'? A similar point is made with respect to assessing the evolutionary potential of fishermen's organizations in the EU:

> 'realization that in terms of federal/national government seeking to influence the development of fishermen's organizations they must adapt their models to suit the historical, political, and social conditions which the current institutions merely reflect; institutions which manifestly fulfil functions that are possibly more important than strictly economic ones'

Through the mechanism of co-management it is anticipated that, amongst other benefits, there would be the development of an harmonious 'architecture' between the fishermen's organizations and others involved in the management process. This would help to identify social objectives and lead to their incorporation in fisheries management strategy. Despite their seductive elegance, concepts of co-management stand in need of a more vigorous critical appraisal in the light of both differing political cultures and the potentially complex structures of the fishing industry. The examples of traditional devolved management systems described in this volume all fit a broad pattern of location in inshore waters, artisanal fisheries and non-industrialized economic structures. The problem is whether devolved management can be developed within more highly centralized political cultures and adapted to the more diverse conditions of the off-shore fisheries.

Property rights

As the crisis in the North Atlantic fisheries has deepened controversial issues with respect to property and use rights have taken centre stage, with numerous instances of conflict between different user groups. In the European Union in particular we have witnessed an uneasy equilibrium between the forces of non-discriminatory 'equal access', supported by the principles of the Common Fisheries Policy and the original Treaty of Rome, and 'right of establishment' which has been used by its supporters to champion restricted access discriminating between groups of users. Several contributors to the volume comment that the concepts of the 'common pond' and 'historically stable quotas' bear a heavy responsibility for the current social crisis in the EU. Thus the supporters of 'right of establishment' have been fighting a rearguard action, apparently losing ground, as the year 2002 looms with the threat of the dismantling of the mechanisms that protected and upheld the *status quo*. Earlier derogations (ostensibly concerned with the protection of fisheries resources within the CFP) that have favoured some groups of users whilst discriminating against others are being dismantled or are under threat of repeal by 2002. Thus the 'spatial' sea territories such as the Irish box no longer exist and we await the outcome of trying to police a large sea territory where much greater fishing activity and potential conflict will take place. Other territorial boundaries, including the 6 to 12 nautical mile limits, are likely to become more 'permeable', resulting in a move towards 'equal access' and diminishing the right of establishment. This can only provoke fresh conflict between user groups and intensify resource management problems.

In a non-spatial context property rights are being manipulated by the development of resource allocation mechanisms based on the privatization of quotas. To date only one North Atlantic state, Iceland, has committed itself fully to a management strategy based on ITQs. But this has been interpreted by some as a 'logical' step in the efforts of regulators to conserve fish stocks, reduce fishing capacity and thus attempt to move away from crisis towards industrial restructuring on the basis of 'efficient' users gaining sway at the expense of their 'non-efficient' colleagues. The social 'fallout' from the application of this approach to EU and other European fisheries has attracted substantial criticism and its implications for sustainable resource management have been questioned. Indeed at least one contributor foresees the 'refeudalization' of the coastal areas of Europe if the current objectives of the CFP are pursued, and the trend of quota privatization continues.

Earlier we have intimated that support for the traditional fisheries institutions combined with devolution of management responsibility are steps in the direction of alleviating the fisheries crisis at least within inshore waters. Therefore we would anticipate that progress in these areas would help with the resolution of property and use rights issues. Indeed would not a satisfactory outcome of the manipulation of property rights in the EU context be in attaining three important objectives: the settlement of resource access disputes on a regional basis; the enhancement of the role of the fishermen in the conservation of the resource; and the reduction of the costs of government control? With respect to the first and second objective several contributors emphasize the value of traditional institutions in creating a sense of 'ownership' and therefore managerial responsibility with respect to the husbandry of the littoral biological resources; and also providing a 'database' of knowledge and skills that can be used for the resource management process. In the Japanese situation it has been demonstrated that territorial use rights, constructed around the concept of the local community, can provide a robust framework for devolved fisheries management even in an aggressively modernizing economy such a Japan. But it would be naive to believe that similar systems could be imposed upon the fisheries of the North Atlantic or even that the maintenance or re-creation of traditional management systems can offer sufficiently sturdy defence against pressures of modernization and over-exploitation. However, given a genuine commitment to the management of inshore fisheries by a co-partnership of local knowledge, experience and expertise on the one hand and state enforcement on the other, it is not unreasonable to anticipate the evolution of appropriate local use rights systems which can mediate the conflicting demands of different resource user groups.

High seas fisheries

While the introduction of exclusive economic zones (EEZs) in the mid-1970s may have brought some sense of order to the fisheries by allocating primary responsibility for their management to the nation state, it certainly did not solve the problems of over-capacity and over-exploitation of the stocks. As the majority of papers in this volume clearly indicate, there is still a major task ahead in first achieving and then maintaining an equilibrium between fishing effort and the sustainable yield of fisheries throughout most of the world's oceans. The creation of EEZs led to a displacement of fishing effort by the distant-water fleets and, as a result, the emergence of new areas of concern for management, principally in the redefined areas of the 'high seas'. New tensions have arisen over the management of so-called straddling (or transboundary) stocks situated on the edge of the continental shelf and in enclaves of international waters within enclosed or semi-enclosed seas, surrounded by the EEZs of several coastal states. Such areas have attracted very heavy fishing effort in recent years, both from the established distant-water fishing nations like Spain, Portugal and Poland and from the new distant-water fleets from east Asian countries such as South Korea and Taiwan. Uncontrolled high seas fishing of straddling stocks can seriously undermine strategies for stock conservation developed by adjacent coastal states in respect of their EEZs. The calming of the water clearly requires that a similar degree of order is brought to bear on the exploitation of fish stocks in international waters.

It is an obvious truth that fish stocks are no respecters of political boundaries. The problem of straddling stocks occurs in almost any part of the world's oceans and will affect most non-sedentary species at some stage of their life cycles, but it has become a critical issue in the management of particular commercial species. For example, stocks of Alaskan pollock in the North Pacific have been affected by high levels of unregulated exploitation of the high seas enclaves in the Bering Sea (the so-called 'donut hole') and the Sea of Okhotsk ('peanut hole'). In the North Atlantic the most controversial issue concerns the transboundary stocks of cod and other bottom-living fish which straddle the 200-mile exclusive fishing zone (EFZ) of eastern Canada, formerly heavily exploited by both Canadian fleets within the EFZ and outside it by the distant-water fleets. These stocks, now seriously depleted, have been subject to a moratorium on fishing within the EFZ. But outside the 200-mile limit, on the Nose and Tail of the Grand Banks, in the area managed by the North Atlantic Fisheries Organization (NAFO), there has been almost unlimited fishing by distant water fleets.

Further problems for high seas management concern both the highly migratory stocks of species like tuna, which may cross several EEZs but which are caught largely in international waters, and the newly exploited deep-water species such as orange roughy and grenadier, located on the continental slopes (Hopper, 1995). The biological characteristics of late maturity, low reproduction rates and longevity render the deep-sea species vulnerable to conventional patterns of exploitation and regulation: several stocks of orange roughy in the South Pacific, for example, are seriously over-exploited and in some cases their biomass reduced to 20% of the original level. Despite considerable biological research, for most deep-water species there is a dearth of reliable stock assessment data on which to base sound management decisions. Such resources must therefore be treated with extreme caution under effective international management.

Management of high seas fisheries is a matter for international law and the responsibility primarily of the United Nations. Unless the issues of straddling and highly migratory stocks are resolved through the establishment of a robust international regime, there is a real risk of high seas chaos resulting from controversial unilateral action to extend coastal state jurisdiction beyond the 200-mile limit. Already Chile has voiced its claim to the *mare presencial* (presential seas) – an area of 2 000 000 square miles beyond the existing EEZ, ostensibly to protect the chub mackerel stock. Likewise the amendment of Canada's Coastal Fisheries Protection Act in 1994 was designed to extend its enforcement powers beyond its own 200-mile limit.

The basic principles for high seas management are set out in the UN Law of the Sea Convention (UNCLOS) adopted in 1982 and which finally came into force in November 1994. With regard to high seas fishing in general, Article 117 lays down the basic obligation that all states have the duty to take, or to co-operate with other states in taking such measures for the respective nationals as may be necessary for the conservation of living resources of the high seas.

Articles 63 and 64 refer to straddling stocks and highly migratory stocks respectively and urge upon the coastal states and others the need for co-operation in seeking to agree appropriate measures. But as Freestone (1995) points out the obligations lack a degree of precision and do not necessarily imply the conclusion of an agreement.

A move to translate these principles into a more substantial binding accord was initiated by Canada during preparatory discussions for the UN Conference on Environment and Development (UNCED). It culminated in the setting up of the special UN Conference on Straddling Fish Stocks and Highly Migratory Fish Stocks in 1993 which met in several sessions before reaching its conclusions in the summer of 1995. The debates opened up many of the old divisions between the coastal states and the distant-water fishing nations and involved a restating of the arguments originally deployed in the run up to the declaration of EEZs in the mid-1970s.

Among the issues which divided the two groups of countries was the implementation of the 'precautionary principle' advocated at the Rio Conference, namely that:

'where there are threats of serious or irreversible damage, lack of full scientific certainty shall not be used as a reason for postponing cost effective measures to prevent environmental degradation.'

(UNCED)

The distant water fishing nations not surprisingly found the precautionary principle, initially devised for pollution control, potentially too restrictive, particularly in view of the increasing uncertainties of fisheries science. But the main concerns were over the views put forward by the coastal states that the need for compatibility between conservation and management measures within and beyond the EEZ effectively required the extension of the coastal state's management regime into the adjacent high seas area.

Even more controversial was the opinion expressed by some coastal states that where the rights of the coastal state and distant-water fleets were in conflict the latter should be subordinate to the former (Hayashi, 1995).

Notwithstanding such deepseated differences of opinion, the conference appears to have brokered a significant international agreement on high seas fisheries management. Subject to ratification, the accord requires regional fisheries organizations such as NAFO to adopt a 'precautionary approach' in setting conservative quotas for straddling stocks. But the accord goes much further by empowering the regional organizations to board, inspect and, in the absence of appropriate action from the flag state, to order a vessel into port (*The Times*, 1995). Thus a measure of enforceable discipline can be brought to bear on fishing outside the 200-mile EEZ. The strength of this agreement will be tested first in its ratification and subsequently in the degree of respect it commands within the international fishing community.

Future social science research in fisheries management

These concluding thoughts would not be complete without some indication of the future directions that social science research in fisheries might take. The role of applied social science research should, in this instance, be to chart the waters of policy change and to fix some navigation aids by which a more secure course can be steered towards the goal of sustainable resource management. There is a clear need for social scientists to redefine their aims and objectives in such a way as to increase their policy relevance. But the greater task is the ecumenical mission to bridge the gap between the perspectives of the 'soft' social science, rooted in the need for flexibility in the face of uncertainty, and the rigidities apparent in the 'scientific laws' of marine biology and fisheries economics. This is not a task for social scientists alone, no matter how broad a church they may claim to represent.

Some of the key topics for investigation jointly with related disciplines are as follows:

- differentiating the values, aspirations and behaviour of fishermen engaged in distinct segments of the fishing industry;
- socio-economic outcomes from alternative property rights systems;
- institutional arrangements for effective fisheries management;
- distributional effects of alternative regulatory systems;
- socio-economic structures relating to the fishing industry at several different spatial levels (family, locality, district, region . . .).

Throughout such an agenda it will be important to confront the erroneous but pervasive image of social science research as introspective and somewhat narrowly focused on the exceptional, thus abstracting the fishing industry from the realities of the wider world. Fisheries should be viewed as an open social system, with internal structures and relationships but also linkages to the economic, social and environmental systems at several points along the continuum from local to global.

Nor should the social scientists neglect the broader issues of the impacts of fisheries policy on the socio-economic structures of the fisher family, the fishing community and the fisheries-dependent region. Here, the thrust of research activity should be directed towards how the resources of the community might be most effectively combined with those of the state to ensure the sustainability of coastal regions and their populations. In particular, consideration should be given to indicating how 'development upwards' notions of economic and social policy might be fostered among resource user groups and integrated with 'development downwards' processes engendered by state intervention.

Underlying this outline agenda for research is a concern to define and elaborate a set of social goals for fisheries policy to set alongside those of resource sustainability and economic efficiency in resource use which emanate from the prevailing bio-economic paradigm. For too long, issues of social equity have been discounted and the social consequences of policy decisions left to others to deal with.

The essential message from social science research is that fisheries management must be more sensitively positioned within the wider economic, social and environmental context if it is to succeed.

References

Crean, K. (1994) Social Objectives and the Common Fisheries Policy. Presented at the Workshop 'An Agenda for Social Science Research in Fisheries Management', The Borschette Centre, Brussels.

Freestone, D. (1995) The Effective Conservation and Management of High Seas Living Resources: Towards a New Regime? *The Canterbury Law Review*, **5** (3) 341–62.

Hayashi, M. (1995) The role of the United Nations in managing the world's fisheries. In *The Peaceful Management of Transboundary Resources* (eds G.H. Blake, W.J. Hildesley, M.A. Pratt, R.J. Ridley & C.H. Schofield), pp. 373–93. Graham & Trotman/Martinus Nijhoff, London/Dordrecht/Boston.

Hopper, A. (ed.) (1995) *Deep-Water Fisheries of the North Atlantic Oceanic Slope.* Kluwer Academic Publishers, Dordrecht/Boston/London.

Kay, J. (1993) *Foundations of Corporate Success – How Business Strategies Add Value.* Oxford University Press, Oxford.

Jentoft, S. (1989) Fisheries co-management: delegating government responsibility to fishermen's organizations. *Marine Policy*, **13** (2), 137–54.

Bibliography

Acheson, J.M. (1988) *The Lobster Gangs of Maine*. Hanover and London University Press, New England.

Adrianou, N. (1987) *Fishing Techniques and Fishing Gears*. Le Pirée, Greece.

Akimichi, T. & Ruddle, K. (1984) The Historical Development of Territorial Rights and Fishery Regulations in Okinawa Inshore Waters. In *Maritime Institutions in the Western Pacific*, (eds K. Ruddle & T. Akimichi), pp.37–88. Senri Ethnological Studies No. 17, National Museum of Ethnology, Osaka.

Alegret, J.L. (1990) Del corporativismo dirigista al pluralismo democrático: last Cofradías de Pescadores de Cataluña. *ERES Serie de Antropologia*, **II** (i), 161–72, Museo Etnografico, Cabildo de Tenerife.

Alegret, J.L. (forthcoming) Co-management and legitimacy in cooperative fishermen's organizations: les Confraries de Pescadors de Catalunya, Spain. In *Proceedings of the World Fisheries Congress* (eds R. Meyer, B. McCay, L. Hushak, E. Buck, & R. Muth). Athens, May 1992.

Anon. (1994) *Forbrugsgoder, – en Erhversøkonomisk Analyse*. Erhvervs-fremmestyrelsen, København.

Árnason, R. (1993) Icelandic fisheries management. In *The Use of Individual Quotas in Fisheries Management*, pp. 123–43. OECD, Paris.

Atlantic Arc Commission (1992) Prospective Study for the Atlantic Regions. *AA Chronicle*, **6**, September, Nantes.

Atlantic Arc Commission (1995) *Proposed Business Plan 1995–99*. Cabinet du Président de la Commission Arc Atlantique, Poitiers, France.

Bailey, C. & Jentoft, S. (1990) Hard choices in fisheries development. *Marine Policy*, **14**, 333–34.

Befu, H. (1980) Political Ecology of Fishing in Japan: Techno-Environmental Impact of Industrialization in the Inland Sea. *Research in Economic Anthropology*, **3**, 323–47.

Berkes, F. (1986) Marine inshore fishery management in Turkey. In *Proceedings of the Conference of Common Property Resource Management*, pp. 63–83. National Research Council, National Academy Press, Washington, DC.

Beverton, R. & Holt, S. (1957) *On the Dynamics of Exploited Fish Populations*. Fisheries Investigations Series 2 (19), Fisheries and Food Department of the Ministry of Agriculture, London.

Boulard, J.C. (1991) L'épopée de la sardine, un siècle d'histoires de pêches. Ouest-France/IFREMER, Rennes/Paris/Brest.

Bourdieu, P. (1972) *Esquisse d'une Théorie de la Pratique.* Droz, Geneva.

Bourdieu, P. (1992) *An Invitation to Reflexive Sociology.* Polity Press, Cambridge.

Boyd, R.O. & Dewees, C.M. (1992) Putting theory into practice: transferable quotas in New Zealand's fisheries. *Society and Natural Resources,* **5**, 179–98.

Bressers, J.Th.A. (1985) *Milieu op de Markt. De Controverse Tussen twee Markt-benaderingen in het Miliubelied.* Kobra, Amsterdam.

Burg & De Raad (1992) *Fiscale Aspecten met Betrekking tot Quota.* Visserijdag, Rotterdam.

Burt, S. (1990) *The Internationalization of European Food Retailers.* Institute for Retail Studies, Glasgow.

Caddy, J.F. & Gulland, J.A. (1985), Historical patterns of fish stocks. *Marine Policy,* **7** (4), 267–78.

Caviedes, C.N. & Fik, T.J. (1992) The Peru–Chile eastern Pacific fisheries and climatic oscillation. In *Climate Variability, Climate Change and Fisheries* (ed M.H. Glantz), pp. 355–75. Cambridge University Press, Cambridge.

Ceccini, P., Catinat, M. & Jacquemin, A. (1988). *Europa '92, Realiseringen af det Indre Marked.* Europakommisionen, Copenhagen.

CEDRE/ECRD (1991) *Etude Prospective des Régions Atlantiques.* Commission of the EC, Directorate General for Regional Policy, Brussels.

Collet, S. (1986) *Poisson et les Hommes.* Rapport pour la Division Sciences Ecologiques, UNESCO, Paris.

Collet, S. (1989) Faire de la parenté, faire du sang: logique et représentation de la chasse à l'espadon. *Etudes Rurales,* **115/116**, 223–50.

Collet, S. (1991) Guerre et pêche: quelle place pour les sociétés de pêcheurs dans le modèle des sociétés de chasseurscueilleurs. *Information sur les Sciences Sociales,* **30** (3), 483–522.

Collet, S. (1993) *Uomini e Pesce: la Caccia al Pesce Spada tra Scilla e Caribdi.* (ed. G. Maimone), Collana Universitatess saggi, Catania.

Collet, S. (1995) Halieutica phoenicia I. Contribution à l'étude de la place des activités dans la culture phénicienne: point de vue d'un non archéologue. *Information sur les Sciences Sociales,* **34** (1), 107–73.

Commission of the EC (1975) *SEC (70) 4503 final.* 22 December, Brussels.

Commission of the EC (1976) *COM (76) 500 final.* 23 September, Brussels.

Commission of the EC (1991) *Europe 2000: Outlook for the Development of the Community's Territory.* COM (91) 452 final, 7 November, Brussels.

Commission of the EC (1994a) *Study of Prospects in the Atlantic Regions.* Official Publications of the EC, Luxembourg.

Commission of the EC (1994b) Notice to the member states laying down guidelines for ... assistance within the framework of a Community initiative concerning the restructuring of the fisheries sector. *Official Journal of the European Commission,* C180/1–5, 1 July.

Cordell, J. (1984) Defending Customary Inshore Sea Rights. In *Maritime Institutions in the Western Pacific,* (eds K. Ruddle & T. Akimichi), pp. 301–26. Senri Ethnological Studies No. 17, National Museum of Ethnology, Osaka.

Cordell, J. (ed.) (1989) *A Sea of Small Boats.* Cultural Survival Inc., Cambridge, Massachusetts.

Cornwall County Council (1994) *South West Pesca Programme: Draft Outline (September 1994).* Economic Development Office, Truro, Cornwall.

Council of the EC (1983a) Regulation (EEC) 170/83 of 25 January establishing a Community system for the conservation and management of fishery resources. *Official Journal of the European Commission*, **26** (124), 1–13.

Crean, K. (1994) Social Objectives and the Common Fisheries Policy. Presented at the Workshop 'An Agenda for Social Science Research in Fisheries Management', The Borschette Centre, Brussels.

Dali, T. & Mørkøre, J. (1983) *Stat og Småborgerligt Fiskeri i den Faerøske Samfundsformation*. Institute of History/Institute of Political Studies, University of Copenhagen.

Dawson, J., Shaw, S., Burt, S. & Rana, J. (1986) *Structural Change and Public Policy in the European Food Industry*. FAST Occasional Papers, No. 103–4, Brussels.

De Boer & Van Keulen (1992) Memorandum inzake fiscale problemen in de Bedrijfstak van visserij. In *Fiscale Aspecten met Betrekking tot Quota*. (eds Burg & De Raad), Visserijdag, Rotterdam.

De Haan, H. (1994) *In the Shadow of the Tree: Kinship, Property and Inheritance Among Farm Families*. Het Spinhuis, Amsterdam.

Delbos, G. & Jorion, P. (1984) *La Transmission des Savoirs*. Maison des Sciences de l'Homme, Paris.

Delbos, G. (1989) De la nature des uns et des autres, à propos du dépeuplement des eaux marines. In *Du Rural à l'Environnement, la question de la Nature Aujourd'hui* (eds N. Mathieu & M. Jollivet), pp. 50–63. L'Harmattan, Paris.

Delbos, G. (1993) L'histoire d'une longue quête, la domestication des poissons plats. In *Cultiver la Mer*, pp. 135–155. Musée Maritime de Saint-Vaast-La-Hougue, Saint-Vaast-La-Hougue.

Dewees, C.M. (1989) Assessment of the implementation of individual transferable quotas in New Zealand's inshore fishery. *North American Journal of Fisheries Management*, **9**, 131–9.

di Natale, A. (1991) Marine mammals' interactions in Scombridae fishery activities: the Mediterranean Sea case. *Fisheries Report*, **449**, FAO, Rome.

di Natale, A. (1992) *Gli attrezzi pelagici derivanti utilizzi per cattura del pesce spada adulto: valutazione comparata della funzionalita, della capacita di cattura, dell'impatto globale e della economia*. Ministero della Marina mercantile, Roma.

Didou, H. (1994) *Réflexions sur l'Avenir des Pêches Bretonnes*. Rapport pour le Conseil Economique et Social Région Bretagne, April 1994, Rennes.

Dilley, R. (ed.) (1992) *Contesting Markets: Analyses of Ideology, Discourse and Practice*, Edinburgh University Press.

Dufour, A.H. (1987) Poser, trîner: deux façons de concevoir la pêche et l'espace. *Bulletin d'Écologie Humaine*, **5**(1), 23–45.

Dufour, A.H. (1990) Leggere e gestire i fondi marini. Due aspetti complementari della pesca nel litoral della Provenza. In *La Cultura del Mare. La Ricerca Folklorica*, **21**, 51–5.

Dufour, A.H. (1993) Les pêches traditionnelles, l'exemple méditerranéen. In *Le Patrimoine Maritime et Fluvial: Actes du Colloque Estuaire 92*, Nantes, Avril 1992, pp. 192–7.

Durrenberger, E.P. & Pálsson, G. (1986) Finding fish: the tactics of Icelandic fishermen. *American Ethnologist*, **13**, 213–29.

Durrenberger, E.P. & Pálsson, G. (1987a) The grass roots and the state: resource management in Icelandic fishing. In *The Question of the Commons: the Culture and Ecology of Communal Resources* (eds B.M. McCay & J.M. Acheson), pp. 370–92. University of Arizona Press, Tucson.

Durrenberger, E.P. & Pálsson, G. (1987b) Ownership at sea: fishing territories and access to sea resources. *American Ethnologist*, **14** (3), 508–22.

Durrenberger, E.P. (1992) *It's All Politics: South Alabama's Seafood Industry.* University of Illinois Press, Chicago.

EF-arbeiðsbólkurin (1991) *Føroyar og EF –Útlit Fyri Samvinnu: Føroya Landsstyrið,* Tórshavn.

Eurofish Report (1991) 1.8.91. Agra Europe Ltd, London.

Eurofish Report (1994) 18.8.94. Agra Europe Ltd, London.

FAO (1992) Marine Fisheries and the Law of the Sea: A Decade of Change. *Fisheries Circular,* No. 853, FAO, Rome.

Ferber, M.A. & Nelson, J.A. (eds) (1993) *Beyond Economic Man: Feminist Theory and Economics.* University of Chicago Press, Chicago.

Fiddes, N. (1991) *Meat, a Natural Symbol.* Routledge, London.

Finlayson, A.C. (1994) *Fishing for Truth: A Sociological Analysis of Northern Cod Assessments from 1977–1990.* St. John's, Institute of Social and Economic Research, Newfoundland.

Fishing News International (1994) 'Spain buys top French trawling fleet'. March 1994, p. 60, Heighway Publications.

Food and Agriculture Organisation (1995) *The State of the World's Fisheries and Aquaculture,* FAO, Rome.

Frangoudes, K. (1993) *Le Rôle des Organizations Professionnelles dans la Gestion des Pêches en Méditerranée, Étude de Cas Concernant la Gréce.* Rapport pour la DG XIV de la CCE.

Frangoudes, K. (1995) Case studies in the Mediterranean Greece and France. Annex to first intermediate report. In *Management of Renewable Resources: institution, Regional Differences and Conflict Avoidance Related to Environmental Policies and Illustrated by Marine Resource Management* EC-DG XII, Environmental Research Programme, contract EV5V-CT94-0386.

Freestone, D. (1995) The Effective Conservation and Management of High Seas Living Resources: Towards a New Regime? *The Canterbury Law Review,* **5** (3), 341–62.

Friis, P. (1993) EFs marked og dets betydning for de nordiske fiskerisamfund. In *Fiskerisamfund-Hvilke Veje?* (ed. G. Gudmundsson), pp. 15–38. NORD 1993:27., København.

Fylling, I., Hanssen, J.I., Sandvin, J. & Størkersen, J.R. (1994) *Vi sto han av!* Report from Nordland Research Institute 16/94, Bodø.

Gatewood, J.B. (1993) Ecology, efficiency, equity, and competitiveness. In *Competitiveness and American Society,* (ed. S.L. Goldman), pp. 123–55. Lehigh University Press, Bethlehem, PA.

Gentles, P. (1993) The place of seafood and retail development in the UK. In *New Markets for Seafood* (ed. W.F.A. Horner & G.A. Rodriguez), pp. 1–25. University of Hull International Fisheries Institute, Hull.

Giddens, A. (1979) *Central Problems in Social Theory.* Macmillan Press, London.

Giddens, A. (1984) *The Constitution of Society.* Polity Press, Cambridge.

Giddens, A. (1990) *The Consequences of Modernity.* Polity Press, Cambridge.

Giddens, A. (1991) *Identity and Self in High Modernity.* Polity Press, Cambridge.

Giddens, A. (1991) *Modernity and Self-Identity, Self and Society in the Late Modern Age.* Stanford University Press, California.

Gilbertsen, N. (1993) Chaos in the commons: salmon and such. *Maritime Anthropological Studies,* **6** (1/2), 74–91.

Giovannoni, V. (1987) *Des Jardiniers de l'eau. Genèse d'une Culture.* Université de Montpellier, rapport dactylographié.

Giovannoni, V. (1988) *Le Mourre Blanc. Du Technique au Social.* Université de Provence, rapport dactylographié.

Glantz, M.H. (ed.) (1992) *Climate Variability, Climate Change and Fisheries.* Cambridge University Press, Cambridge.

Gordon, H.S. (1954) The economic theory of a common property resource: the fishery. *Journal of Political Economy,* **62**, 124–42.

Graham, M. (1935) Modern theory of exploiting a fishery and application to North Sea trawling. *Journal du Conseil International pour l'Exploration de la Mer,* **10**, 264–74.

Granovetter, M. (1985) Economic action and social structure: the problem of embeddedness. *American Journal of Sociology,* **91** (3), 481–510.

Granovetter, M. (1992) Economic institutions as social constructions: a framework for analysis. *Acta Sociologica,* **35**, 3–11.

Gudeman, S. (1986) *Economics as Culture: Models and Metaphors of Livelihood.* Routledge & Kegan Paul, London.

Gudeman, S. (1992) Remodeling the house of economics: culture and innovation. *American Ethnologist,* **19** (1), 139–52.

Gulland, J.A. (1987) The management of North Sea fisheries: looking towards the 21st century. *Marine Policy,* **11**, 259–71.

Gurevich, A. (1992) *Historical Anthropology of the Middle Ages,* (ed J. Howlett), Polity Press, Oxford.

Habernas, J. (1978) *Raison et Légitimité.* Payot, Paris.

Hanna, S.S. (1990) The eighteenth century English commons: a model for ocean management. *Ocean & Shoreline Management,* **14**, 155–72.

Hannesson, R. (1993) *Bio-economic Analysis of Fisheries.* Blackwell Scientific Publications, Oxford.

Hardin, G. (1968) The Tragedy of the Commons. *Science,* **162**, 1243–8.

Hayashi, M. (1995) The role of the United Nations in managing the world's fisheries. In *The Peaceful Management of Transboundary Resources* (eds G.H. Blake, W.J. Hildesley, M.A. Pratt, R.J. Ridley & C.H. Schofield), pp. 373–93. Graham & Trotman/Martinus Nijhoff, London/Dordrecht/Boston.

Helgasson, A. (1995) *The Lords of the Sea and the Morality of Exchange: the Social Context of ITQ Management in the Icelandic Fisheries.* M.A. thesis, University of Iceland.

Hersoug, B. (ed.) (1992) *Fiskerinœringens Hovedtrekk – Landsanalyser av Danmark, Fœrøerne, Grønland, Island og Norge.* Nord 1992:30, Copenhagen.

Hilborn, R. & Sibert, J. (1988) Adaptive management of developing fisheries. *Marine Policy,* **12**, 112–21.

Hobbes, T. (1968) *Leviathan* (ed. C. Macpherson), Penguin Books, Harmondsworth, UK.

Hoefnagel, E. (1993) Kettingreacties in het visserijbeleid – via onbedoelde naar beoogde beleidseffecten. *Facta,* **4**, 2–7.

Høgnesen, O.W. (1988) Samfelagslig Lysing av útróðrarstœrarstœttini v.m. Lecture at 'Útróðrarstevnan í Klaksvík', Conference for the Inshore Fishermen, Klaksvík.

Holden, M. (1994) *The Common Fisheries Policy*. Blackwell Science Ltd, Oxford.

Holm, P. (1995a) Fisheries management and the domestication of nature, paper presented to the Fifth Annual Common Property Conference: Reinventing the Commons. Bodø, Norway, 24–28 May, 1995.

Holm, P. (1995b) Productionism, resource management and co-management: phases in the development of Norwegian fisheries. Paper presented to the XVI European congress on Rural Sociology, Prague, 31 July–4 August, 1995.

Hopper, A. (Ed.) (1995) *Deep-Water Fisheries of the North Atlantic Oceanic Slope*. Kluwer Academic Publishers, Dordrecht/Boston/London.

Howell, D.L. (1995) *Capitalism from Within. Economy, Society, and the State in a Japanese fishery*. University of California Press, Berkeley.

Jentoft, S. (1989) Fisheries co-management: delegating government responsibility to fishermen's organizations. *Marine Policy*, **13** (2), 137–54.

Johannesen, K. (1980) *Faerøsk Fiskeri – og Markedspolitik i 70'erne*. Institute of Political Studies, University of Århus.

Jørgensen, N. (1989) *Jagten på den Forsvundne Fordel. EFs Indre Marked og Dansk Landbrugs og Foraedlingsindustris Mulighed*. Handelshøjskole Syd. Kolding.

Jorion, P. (1982) All brother crews in the North Atlantic. *The Canadian Review of Sociology and Anthropology*, **19** (4), 513–26.

Jorion, P. (1983) *Les Pêcheurs d'Houat, Anthropologie Économique*. Herman, Paris. Judgement of the Court of Justice of the EC (1991) Case 221/89, 25 July, *Official Journal of the European Commission*, C220 23.8.91:5.

Kalland, A. (1981) *Shingu: A Study of a Japanese Fishing Community*. Curzon Press, London.

Kalland, A. (1984) Sea Tenure in Tokugawa Japan: The Case of Fukuoka Domain. In *Maritime Institutions in the Western Pacific*, (eds K. Ruddle & T. Akimichi), pp. 11–36. Senri Ethnological Studies No. 17, National Museum of Ethnology, Osaka.

Kalland, A. (1987) In Search of the Abalone: The History of the Ama of Northern Kyushu, Japan. In *Seinan chiiki no shiteki tenkai*, pp. 23–33. Shibunkaku Shuppan, Kyoto.

Kalland, A. (1994) *Fishing Villages in Tokugawa, Japan*. University of Hawaii Press, Curzon Press, London/Honolulu.

Kay, J. (1993) *Foundations of Corporate Success – How Business Strategies Add Value*. Oxford University Press, Oxford.

Kottak, C.P. (1992) *Assault on Paradise: Social Change in a Brazilian Village*. 2nd edn. McGraw-Hill, New York.

Kunneke, R.W. (1992) De verdeling van eigendomsrechten als bestuurkundig vraagstuk. *Bestuurskunde*, **4**.

Lave, J. & Wenger, E. (1991) *situated Learning: Legitimate Peripheral Participation*. Cambridge University Press, Cambridge.

Lave, J. (1988) *Cognition in Practice: Mind, Mathematics and Culture in Everyday Life*. Cambridge University Press, Cambridge.

Le Bihan, A. (1958) Les préfets du Finistère. In *Bulletin de la Société Archéologique du Finistère, Quimper, France*.

Lindblad, I. *et al.* (1984) *Politik i Norden: En Jämförande Översikt*. Liber Förlag, Stockholm.

LNV (1992) Vissen naar evenwicht. *Beleidsvoornemen. Structuurnota Zee – en Kustvisserij.* Ministerie van Landbouw, Natuurbeheer en Visserij.

Mariussen, A., Muller, H. & Høydahl, E. (1994) *En Fortelling om to Kystregioner.* Report from Nordland Research Institute Rapport 13/94, Bodø.

Mariussen, A., Onsager, K. & Tønnesen, S. (1990) *Fiskerikrise – Virkninger og Omstilling i Kystsamfunn.* NIBR notat 134, NIBR, Oslo.

McCay, B. & Acheson, J.M. (eds) (1987) *The Question of the Commons. The Culture and Ecology of Communal Resources.* University of Arizona Press, Tucson.

McCay, B.M. & Creed, C.F. (1990) Social structure and debates on fisheries management in the Atlantic surf claim fishery. *Ocean and Shoreline Management,* **13**, 199–229.

McCay, B.M. & Creed, C.F. (1993) *Social Impact of ITQs in the Sea Clam Fisheries.* Rutgers University (Mimeo).

McCloskey, D. (1985) *The Rhetoric of Economics.* University of Wisconsin Press, Madison, WI.

McEvoy, A.F. (1988) Toward an interactive theory of nature and culture: ecology, production, and cognition in the California fishing industry. In *The Ends of the Earth: Perspectives on Modern Environmental History* (ed. D. Worster), pp. 211–229. Cambridge University Press, Cambridge.

McGoodwin, J.R. (1990) *Crisis in the World's Fisheries.* Stanford University Press, Stanford.

McKean, M.A. (1981) *Environmental Protest and Citizen Politics in Japan.* University of California Press, Berkeley.

Merchant, C. (1980) *The Death of Nature: Women, Ecology and the Scientific Revolution.* Harper & Row, San Francisco.

Merchant Shipping Act (1988) *UK Statutes in Force.* HMSO, London.

Ministry of Fisheries (1993) *Annual Report of the Ministry of Fisheries.* Ministry of Fisheries, Tokyo.

Ministry of the Government Presidency (1970) *Fisheries Code.* Decree 420/1970, Ethniko Typographico.

Mitrany (1975) *The Functional Theory of Politics.* Robertson, London.

Mollo, P. (1992) *La pêche sur L'île de Houat.* Mémoire de maitrise de l'UF Anthropologie, Ethnologie et Sciences des Religions, Université de Paris VII.

Montfort, M.C. (1993) The French Market for seafood – recent developments and prospects. In *New Markets for Seafood* (ed. W.F.A. Horner & G.A. Rodriguez), pp. 26–34. University of Hull International Fisheries Institute, Hull.

Mørkøre, J. (1991a) Et kroporativt forvaltningsregimes sammenbrud – erfaringer fra det faerøske fiskeri i nationalt farvand. In *Fiskerireguleringer,* pp. 44–82. Nordiske Seminar og Arbejdsrapporter, Copenhagen.

Mørkøre, J. (1991b) Class interests and nationalism in Faroese politics. *North Atlantic Studies,* **1**, 57–67.

Morvan, Y. (1989) 'Le renforcement de l'Arc Atlantique: une opportunité pour les acteurs économiques de l'Ouest européen'. Paper presented at the Assemblée Générale de l'Association Ouest-Atlantique, Brest 16.11.89.

Motais, M. (1981) *La Pêche Française en Méditerranée.* Mémoire pour le DESS de droit maritime, Aix-Marseille.

Neher, P.A., Árnason, R. & Mollett, N. (eds) (1989) *Rights Based Fishing.* Kluwer Academic Publishers, Dordrecht.

Nentjes, A. (1988) De economie van het mestoverschot. *Tijdschrift voor Milieukunde*, **3** (5), 159–64.

Neystabø, K. (1984) *Faerøerne og EF*. Egio forlag, Tórshavn.

Ninohei, T. (1981) *Fisheries Exploitation in the Meiji Era*. Heibonsha, Tokyo.

North, D.C. (1981) *Structure and Change in Economic History*. Norton & Co., New York.

North, D.C. (1991) *Institutions, Institutional Change and Economic Performance*. Cambridge University Press, New York.

North, E.C. (1990) *Institutions, Institutional Change and Economic Performance*. Cambridge University Press, Cambridge.

Oka, N., Watanabe, H. & Hasegawa, A. (1962) The Economic Effects of the Regulation of the Trawl Fisheries of Japan. In *Economic Effects of Fishery Regulations* (ed. R. Hamlisch), pp. 171–208. *Fisheries Report 5*, FAO, Rome.

Okada, O. (1979) Japanese Fisher people's Fight Against Nuclear Power Plants. *AMPO*, **11** (1), 38–45.

Okun, A.M. (1975) *Equality and Efficiency: the Big Trade-off*. The Brookings Institute, Washington, DC.

Olafsson, Á. (1987) Uttanríkistiourskifti Føroya/Faerøernes Undenrigsrelationer/The foreign relations of the Faroe Islands. In *Fiskivinnuárbókin '87 Fiskeriårbogen '87/Fisheries Yearbook '87*, pp. 33–36. S.J. Repro, Tórshavn.

Olafsson, A. (1989) Føroyar og EF/Faerøerne og EF/The Faroe Islands and the EEC. In *Fiskivinnuárbókin '89/Fiskeriårbogen '89/Fisheries Yearbook '89*, pp. 53–63. S.J. Repro, Tórshavn.

Onsager, K. & Eikeland, S. (1992) *Ny Giv i Nordnorsk Kystindustri*. NIBR rapport 8, NIBR, Oslo.

Onsager, K. (1991) *På Rett Kjøl i Åpent Farvann?* Regionale trender nr 2.

Ostrom, E. (1991) *Governing the Commons: the Evolution of Institutions for Collective Action*. Cambridge University Press, New York.

Pálsson, G. (1991) *Coastal Economies, Cultural Accounts: Human Ecology and Icelandic Discourse*. Manchester University Press, Manchester.

Pálsson, G. (1994). Enskilment at sea. *Man*, **29** (4), 901–27.

Pálsson, G. & Durrenberger, E.P. (1990) Systems of production and social discourse: the skipper effect revisited. *American Anthropologist*, **92**, 130–141.

Panayotou, T. (1982) Management Concepts for Small Scale Fisheries: Economic and Social Aspects. *Fisheries Technical Paper 228*, FAO, Rome.

Panorama (1994) *Yearbook from the EU about the State of Industries in the Community*. Brussels.

Peeters, M. (1992) Marktconform Milieurecht? Een Rechtsvergelijkende Studie naar de Verhandelbaarheid van Vervuilingsrechten. Tjeenk Willink Zwolle.

Petersen, C. (1894) On the biology of our flatfishes and on the decrease of our flatfish fisheries. *Report of the Danish Biological Station*, **4**, 1–85.

Pinkerton, E. (1989) *Co-operative Management of Local Fisheries*. University of British Columbia, Vancouver.

Proutière-Mouillon, G. (1993) *La CEE et le marché des produits de la mer; mécanismes juridiques*. Rapport pour le FIOM et le FROM-Bretagne, October 1993, Paris/Brest.

Ritzer, G. (1993) *The Macdonaldization of Society*. Pine Forge Press, London.

Rokkan, S. & Urwin, D.W. (1982) *The Politics of Territorial Identity: Studies in*

European Regionalism. Sage, London.

Rokkan, S. & Urwin, D.W. (1983) *Economy, Territory, Identity: Politics of West European Peripheries.* Sage, London.

Roseberry, W. (1991) Potatoes, sacks, and enclosures in Early Modern England. In *Golden Ages, Dark Ages: Imagining the Past in Anthropology and History.* University of California Press, Berkeley.

Ruddle, K. (1985) The Continuity of Traditional Practices: The Case of Japanese Coastal Fisheries. In *The Traditional Knowledge and Management in Coastal Systems in Asia and the Pacific,* (eds K. Ruddle & R.E. Johannes), pp. 158–79. UNESCO, Jakarta.

Ruddle, K. (1987) Administration and conflict management in Japanese coastal fisheries. *FAO Fisheries Technical Paper 273.* FAO, Rome.

Russell, E. (1931) Some theoretical considerations on the overfishing problem. *Journal du Conseil International pour l'Exploration de la Mer,* **6**, 3–20.

Sagdahl, B. (1992) *Co-management, a Common Denominator for a Variety of Organizational Forms.* Forms from Nordland Research Institute, Bodø.

Salz, P. (1991) *De Europese Atlantische Visserij: Structuur, Economische Situatie en Beleeid.* Onderzoekverslag, **85**, LEI-DLO.

Sandberg, A. (1991) *Fish for All – CPR-Problems in North Atlantic Environments.* NF-Notat no. 1104/91, Bodø.

Sandberg, A. (1993) *The Analytical Importance of Property Rights to Northern Resources.* LOS i NORD-NORGE Notat no. 18, Tromsø.

Schumpeter, J. (1934) *The Theory of Economic Development.* Harvard University Press, Harvard, Massachusetts.

Scott, A.D. (1989) Conceptual origins of rights based fishing. In *Rights Based Fishing* (eds P.A. Neher, R. Arnason & N. Mollett), pp. 11–38. Kluwer Academic Publishers, Dordrecht.

Seierstad, S. (1985) Arbeid som rettighetsgivende status i et kapitalistisk arbeidsliv. *Tidsskrift for Arbeiderbevegelsens Historie 1.*

Sen, A. (1973) *On Economic Inequality.* Oxford University Press, Delhi.

Short, K.M. (1989) Self-management of Fishing Rights by Japanese Cooperative Associations: A Case Study from Hokkaido. In *A Sea of Small Boats,* (ed. J.C. Cordell), pp. 371–87. Cultural Survival Inc., Cambridge, Massachusetts.

Smith, M.E. (1990) Chaos in fisheries management. *Maritime Anthropological Studies,* **3** (2), 1–13.

Spencer, R.F. (1959) *The North Alaskan Eskimo: a Study in Ecology and Society.* Bureau of American Ethnology, Bulletin 171, Washington DC.

Spinoza, B. (1965) Traité theologico-politique. In *Oeuvres* (eds. Garnier & Flammarion), T.2., Paris.

Spinoza, B. (1966) Traité politique In *Oeuvres* (eds. Garnier & Flammarion), T.4., Paris.

Symes, D. & Crean, K. (1994) Social Issues and the Socio-Economic Paradigm in Fisheries Management. Theme paper for workshop entitled 'An Agenda for Social Science Research in Fisheries Management', The Borschette Centre, Brussels.

Symes, D. (1994) The European Pond: Who Actually Manages the Fisheries? Paper presented at the International Geographical Union: Commission on Marine Geography Conference on Public Policy and Ocean Development, St Mary's University, Halifax, Nova Scotia, Canada, 12–16 May 1994.

Townsend, R.E. (1995) Fisheries self-governance: corporate or co-operative structures. *Marine Policy,* **19** (1), 39–45.

Treaty establishing the European Economic Community (1987). In *Treaties Establishing the European Communities.* Office for Official Publications of the EC, Luxembourg.

Veggeland, N. (1992) *Regionalpolitikkken Udfordres.* NordREFO, Copenhagen.

Vestergaard, T.A. (1990) The fishermen and the nation: the identity of a Danish occupational group. *Maritime Anthropological Studies,* **3** (2), 14–34.

Vestergaard, T.A. (1991a) Are artisanal fisheries backward? Artisanal fisheries in modern society, the example of Denmark. In *La Recherche Face à la Pêche Artisanale. Symp. Int.* (eds J.R. Durand, J. Lemoalle & J. Weber), tome II, pp. 781–8. ORSTOM-IFREMER, Montpellier, France.

Vestergaard, T.A. (1991b) Living with pound nets: diffusion, invention and implications of a technology. *Folk Journal of the Danish Ethnographic Society,* **33**, 149–67.

Vestergaard, T.A. (1994) Catch regulation and Danish fisheries culture. *North Atlantic Studies,* **3** (2), 25–31.

Vince, A. (1966) *Entre Loire et Vilaine, Étude de Géographie Humaine.* Thèse de géographie humaine, Université de Poitiers, Faculté des Lettres.

Wang, Z. (1968) Faerøsk politik i nyere tid. In *Faeringer-Fraender: Sprog, Historie, Politik og Økonomi,* (ed. A. Ølgaard), pp. 76–110. Gyldendal, Copenhagen.

Wang, Z. (1988) *Stjórnmálafrøoi.* Futura, Hoyvík.

Weber, J. (1992) Environment: développement et propriété. Une approche épistémologique. In *Gestion de l'environnement.* Ethique et société, pp. 268–301. Fidès, Québec.

Weber, J. (1993) *Le Rôle des Organizations Professionnelles dans la Gestion des Pêches en Méditerranéedes: Synthèse.* ASCA-CCE DG XIV, contrat XIV–12/MED/91/010.

Weber, P. (1994) *Net Loss: Fish, Jobs and the Marine Environment.* Worldwatch Paper 120, Worldwatch Institute, Washington.

White, D.R.M. (1989) Knocking 'em dead: Alabama shrimp boats and the 'fleet effect'. In *Marine Resource Utilization: A Conference on Social Science Issues* (eds J.S. Thomas, L. Maril, & E.P. Durrenberger), Univ. of South Alabama College of Arts and Sciences Publication Vol. 1 & the Mississippi–Alabama Sea Grant Consortium, Mobile.

Wigen, K. (1989) Shifting Control of Japan's Coastal Waters. In *A Sea of Small Boats,* (ed. J.C. Cordell), pp. 388–410. Cultural Survival Inc., Cambridge, Massachusetts.

Williams, R. (1976) *Keywords: a Vocabulary of Culture and Society.* Fontana, Glasgow.

Williamson, O.E. (1975) *Markets and Hierarchies.* Free Press, New York.

Wilson, J.A. (1982) The economical management of multispecies fisheries. *Land Economics,* **58** (4), 417–34.

Wilson, J.A. (1990) Fishing for knowledge. *Land Economics,* **66**, 12–29.

Wilson, J.A. & Kleban, P. (1992) Practical implication of chaos in fisheries. *Maritime Anthropological Studies,* **5** (1), 67–75.

Wilson, J.A., French, J., Kleban, P., McCay, S., Ray, N., & Townsend, R. (1991) Economic and biological benefits of interspecies switching in a simulated chaotic fishery. In *La recherche Face à la Pêche Artisanale. Symp. Int.* (eds J.R. Durand, J. Lemoalle & J. Weber), tome II, pp. 789–802. ORSTROM-IFREMER, Montpellier, France.

Wilson, J.A., Acheson, J.M., Metcalfe, M. & Kleban, P. (1994) Chaos, complexity and community management of fisheries. *Marine Policy,* **18** (4), 291–305.

Wise, M. & Gibb, R. (1993) *Single Market to Social Europe.* Longman, London.

Wise, M. (1984) *The Common Fisheries Policy of the European Community.* Methuen, London.

Yven, A. (1991) *L'Endettement des Familles de Pêcheurs.* Mémoire de Diplôme Supérieur de l'Institut Régional du Travail Social, Rennes.

Index